149
Advances in Polymer Science

Springer-Verlag Berlin Heidelberg GmbH

Biomedical Applications Polymer Blends

With contributions by
G.C. Eastmond, H. Höcker, D. Klee

 Springer

This series presents critical reviews of the present and future trends in polymer and biopolymer science including chemistry, physical chemistry, physics and materials science. It is addressed to all scientists at universities and in industry who wish to keep abreast of advances in the topics covered.

As a rule, contributions are specially commissioned. The editors and publishers will, however, always be pleased to receive suggestions and supplementary information. Papers are accepted for „Advances in Polymer Science" in English.

In references Advances in Polymer Science is abbreviated Adv. Polym. Sci. and is cited as a journal.

ISSN 0065-3195

ISBN 978-3-662-15632-2   ISBN 978-3-540-48838-5 (eBook)
DOI 10.1007/978-3-540-48838-5

Library of Congress Catalog Card Number 61642

© Springer-Verlag Berlin Heidelberg 1999

Originally published by Springer-Verlag Berlin Heidelberg New York in 1999.
Softcover reprint of the hardcover 1st edition 1999

Typesetting: Data conversion by MEDIO, Berlin
Cover: E. Kirchner, Heidelberg
SPIN: 10702256   02/3020 - 5 4 3 2 1 0 - Printed on acid-free paper

Editorial Board

Prof. Samuel I. Stupp
Department of Measurement Materials Science
and Engineering
Northwestern University
2225 North Campus Drive
Evanston, IL 60208-3113, USA
E-mail: s-stupp@nwu.edu

Prof. Ulrich W. Suter
Department of Materials
Institute of Polymers
ETZ,CNB E92
CH-8092 Zürich, Switzerland
E-mail: suter@ifp.mat.ethz.ch

Prof. Edwin L. Thomas
Room 13-5094
Materials Science and Engineering
Massachusetts Institute of Technology
Cambridge, MA 02139, USA
E-mail. thomas@uzi.mit.edu

Prof. Gerhard Wegner
Max-Planck-Institut für Polymerforschung
Ackermannweg 10
Postfach 3148
D-55128 Mainz, FRG
E-mail: wegner@mpip-mainz.mpg.de

Prof. Robert J. Young
Manchester Materials Science Centre
University of Manchester and UMIST
Grosvenor Street
Manchester M1 7HS, UK
E-mail: robert.young@umist.ac.uk

Contents

Advances in Polymer Science
Now Also Available Electronically

For all customers with a standing order for Advances in Polymer Science we offer the electronic form via LINK free of charge. Please contact your librarian who can receive a password for free access to the full articles. By registration at:

http://link.springer.de/series/aps/reg_form.htm

If you do not have a standing order you can nevertheless browse through the table of contents of the volumes and the abstracts of each article at:

http://link.springer.de/series/aps/

There you will find also information about the

– Editorial Bord
– Aims and Scope
– Instructions for Authors

Polymers for Biomedical Applications: Improvement of the Interface Compatibility

Doris Klee, Hartwig Höcker

Department of Textile Chemistry and Macromolecular Chemistry, RWTH Aachen, Veltmanplatz 8, D-52062 Aachen, Germany
e-mail: klee@dwi.rwth-aachen.de

The true aim of biomaterials research is to create implant surfaces which interact actively with the biological system and provoke exactly the same reactions as the corporal tissues do. The improvement in the interface compatibility of polymers selected for implantation by directed surface modification is an important contribution to biomaterial development. Different polymer properties are adjusted and characterized independently of the carrier polymer by means of introduction of modern surface analytical methods and surface techniques. In addition, the interactions between the modified polymer surface and the biological system are measured. In this way, the hydrophilization of a polyurethane (Tecoflex™) and a poly(ether sulfone) by plasma induced graftcopolymerization of hydrogels like poly(hydroxyethyl methacrylate) leads to improved blood compatibility. Functionalization by means of SO_2 plasma treatment of medical grade poly(vinyl chloride) increases the adsorption of the basal membrane protein fibronectin, which correlates with an improvement in cell growth. A suitable interface for an improved cell growth of human vascular endothelial cells as well as for cornea endothelial cells has been created by immobilization of the cell adhesion mediator fibronectin using bifunctional spacer molecules at several carrier polymer surfaces like smooth poly(vinyl chloride), modified polyurethane, Tecoflex™ and poly(dimethyl siloxane).

Keywords. Biomaterials, Interfacial compatibility, Surface modification

List of Abbreviations

AAc	acrylic acid
ABHG	pentanedioic acid mono-4-(acryloyloxy)butyl ester
ABTS	diammonium-2,2'-azinobis(3-ethylbenzothiazolinesulfonate)
AFM	atomic force microscopy
4-amino-TEMPO	4-amino-2,2,6,6-tetramethylpiperidin-*N*-oxyl
aPTT	activated partial thrombin time
Ar-HEMA-PES	argon plasma treated and HEMA-grafted poly(ether sulfone)
Ar-PW-PES	argon plasma treated poly(ether sulfone) stored in water
ATR	attenuated total reflectance
DCC	dicyclohexylcarbodiimide
ECM	extracellular matrix
EDC	1-ethyl-3-(3-dimethylaminopropyl)carbodiimide

ELISA	enzyme-linked immunosorbent assay
ESR	electron spin resonance
EVA	poly(ethene-*co*-vinyl acetate)
EVACO	poly(ethene-*co*-vinyl acetate-*co*-carbon monoxide)
F	force
Fn	fibronectin
GMA	glycidyl methacrylate
GRGDS	pentapeptide consisting of glycine (G), arginine (*R*), aspartic acid (D) and serine (*S*)
h	Planck's constant
HBA	hydroxybutylacrylate
HDI	hexamethylene diisocyanate
HEMA	2-hydroxyethylmethacrylate
HUVEC	human umbilical vein endothelial cells
IR	infrared spectroscopy
MDA	4,4'-diaminodiphenylmethane
MDI	4,4'-diphenylmethane diisocyanate
p	perimeter
P(ABHG)	poly[pentanedioic acid mono-4-(acryloyloxy)butyl ester]
PAAc	poly(acrylic acid)
PAS	photoacoustic spectroscopy
PCU	poly(carbonate urethane)
PCU/EVA	polymer blend consisting of poly(carbonate urethane) and poly(ethene-*co*-vinyl acetate)
PCU/HBA	4-hydroxybutylacrylate modified poly(carbonate urethane)
PCU/PVA	polymer blend consisting of poly(carbonate urethane) and poly(vinyl alcohol)
PDMS	poly(dimethylsiloxane)
PEO	poly(ethylene oxide)
PES	poly(ether sulfone)
PEU/VA	poly(ether urethane) grafted with vinyl acetate
PHBA	poly(4-hydroxybutyl acrylate)
PHBA-COOH	poly(4-hydroxybutyl acrylate) modified with glutaric anhydride
PHBA-COOR	PHBA-COOH after coupling with 4-amino-TEMPO
PHEMA	poly(2-hydroxyethylmethacrylate)
PP/EVA	poly(ethylene-*co*-vinyl acetate)
PPE	poly(ethylene-*co*-propene)
PPE/EVA	blends of poly(ethylene-*co*-propene)/poly(ethylene-*co*-vinyl acetate)
PTT	partial thromboplastin time
PUR	polyurethane
PVA	poly(vinyl alcohol)
PVA-COOMe	model surface poly(vinyl alcohol) reacted with 4-isocyanato-methyl butanoate

PVA-COONa	saponified PVA-COOMe
PVAOCONHPh	model surface poly(vinyl alcohol) reacted with phenylisocyanate
PVC	poly(vinyl chloride)
PVC/EVACO	poly(ethylene-co-vinyl acetate)-graft-vinyl chloride
RGD	tripeptide consisting of arginine (R), glycine (G) and aspartic acid (D)
RGDV	oligopeptide consisting of arginine (R), glycine (G), aspartic acid (D) and valine (V)
SEM	scanning electron microscopy
SIMS	secondary ion mass spectrometry
TCPS	tissue culture polystyrene
Tecoflex/COOR	Tecoflex™ with immobilized 4-amino-TEMPO
TEM	transmission electron microscopy
TFE	trifluoroethylamine
TOF-SIMS	time-of-flight secondary ion mass spectrometry
v	velocity of the secondary ions
ν	frequency
XPS	X-ray photoelectron spectroscopy
ζ-potential	streaming potential
θ_a	contact angle (advancing)
θr	contact angle (receding)
γ_l	surface tension of liquid
γ_{sv}^d	dispersive portion of surface tension
γ_{sv}^p	polar portion of surface tension
γ_s^d	dispersive portion of surface tension of solid
γ_s^p	polar portion of surface tension of solid
γ_{sv}	interfacial tension at the interface solid/air
γ_{sw}	interfacial tension at the interface solid/water

1
Introduction

1.1
Biomaterials

Together with the advances made in structural and functional substances over the last few decades there has also been an increasing number of developments in materials for use in biomedical technology. Approximately 40 years ago the first synthetic materials were successfully employed in the saving and prolonging of human life [1]. Among these were the first artificial heart valves, pacemakers, vascular grafts and kidney dialysis. In the following years, advances in materials engineering made possible the use of orthopedic devices such as knee

Table 1. Biomaterials and their applications

Material class	Material	Application
Metals and alloys	Steel	Fracture correction
		Bone/articular replacement
	Titanium	Dental replacement
		Pace-makers
		Encystem
	Gold alloys	Dental implants
	Silver	Antibacterial
Ceramics and Glasses	Calcium phosphate	Bone regeneration
	Bioactive glass	Bone replacement
	Porcelain	Dentures
Polymers	Polyethylene	Articular replacement
	Polypropylene	Suture materials
	Polytetrafluoroethylene	Vascular grafts
	Polyester	Vascular grafts,
		Resorbable systems
	Polyurethanes	Blood contact devices
	Polyvinyl chloride	Tubes and bags
	Polymethylmethacrylate	Intraocular lenses
	Polyacrylate	Dental implants
	Silicon	Soft tissue replacement
		Ophthalmology
	Hydrogels	Ophthalmology

and hip joint replacements as well as intraocular lenses in the treatment of cataract patients. During this period the choice of materials used was more or less a process of trial and error [2].

At the beginning of the 1980s the concept "*biomaterial*" was defined by the NIH Consensus Development Conference on the Clinical Applications of Biomaterials (1982) [3] as *any substance, other than a drug, or combination of substances, synthetic or natural in origin, which can be used for any period of time, as a whole or as a part of a system which treats, augments, or replaces any tissue, organ or function of the body*. Though an essential observation, Anderson further defined this concept by stating that a biomaterial is a synthetic or modified natural material that interacts with parts of the body [3].

Numerous clinical justifications exist for the employment of biomaterials. Materials are needed for the replacement of tissues that have either been damaged or destroyed through pathological processes. The performance of the biomaterial utilized in the implant must fulfill those functions of the body parts being replaced, e.g. cardiac valves or intraocular lenses. Furthermore, the use of

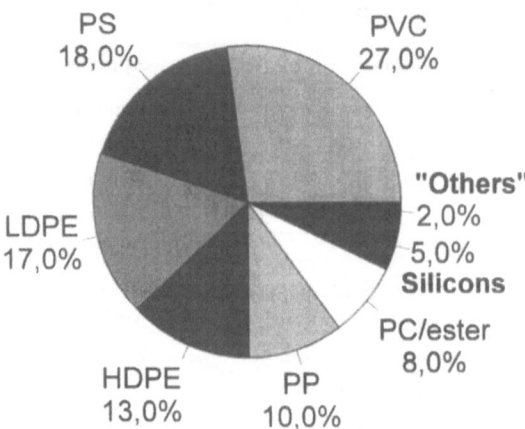

Fig. 1. Synthetic polymers in medicine

biomaterials is required for the removal of congenital defects such as cardiac disorders and for corrective cosmetic surgery. In the treatment of wounds the application of biomaterials could be in the form of suture materials, artificial ligaments and bone fixators. Additional uses of biomaterials are coatings for sensors and pacemakers as well as drug delivery systems within the body. Despite numerous application possibilities the characteristics of biomaterials are somewhat less than desired and should only be employed when there are no available human transplant materials [4]. Biomaterials currently in use in implantation include metals, ceramics, glasses, polymers and composites. Table 1 lists some examples of biomedical materials and their areas of application.

Due to the increase in human life expectancy, an increasing and constant demand for biomaterials has developed. In 1993 in the United States alone, approximately 1.4 million cataract patients were treated and fitted with intraocular lenses. In the area of cardiovascular surgery 250,000 pacemakers and 120,000 heart valves are implanted yearly worldwide [5].

In recent years biomaterials have gained increased importance through object-oriented synthesis, blends, and modifications that produce tailor-made characteristics for the areas where these materials are to be used. Despite all the progress that has been made, the structure of the tailor-made polymers is relatively simple in comparison to that of the complex cellular structure of the tissues being replaced. Most polymer implants are produced from standard polymer substances; 85% of all implants are produced from vinyl polymers. Silicone, polycarbonate, and polyesters comprise 13% of the world market and "other" specially optimized biomaterials are used in 2% of all implants [6] (see Fig. 1).

The most important characteristics required for biomaterials in their numerous operational areas are *biofunctionality* and *biocompatibility*. In most cases the functionality is satisfied through the various mechanical characteristics of currently available materials. Due to the high production standards in materials engineering, it is possible to produce high quality products of suitable design. Nevertheless, these products must retain their functions in an aggressive environment effectively and safely over the desired period of time without irritation of the surrounding tissue by either mechanical action or possible degradation products. This is ensured only when the biomaterial is biocompatible. As a result of the complex interaction between the implant and the tissue generally the expectation of an unsatisfactory biocompatibility of the implant is high. Only a better understanding of the interaction between biomaterials and the biosystem will lead to the development of suitable biomaterials and the successful use of implants [7].

Implant materials, e.g. silicone breast implants, have developed complications which has led to the revoking of licenses by the Food and Drug Administration (FDA). In the US, damage claims against companies manufacturing implants are settled by the Justice Department. Dow Chemicals recently removed its blood contact implant polyetherurethane Pellethane from the market, because in the decomposition of this polyetherurethane in the body the carcinogenic metabolite 4,4'-diaminodiphenylmethane (MDA) could be released [8]. As a consequence of these developments new or improved biomaterials are needed that meet the high standards of the FDA and the medical products laws.

1.2
Biocompatibility

In earlier definitions of material and organism biocompatibility was equated with inertia. The so-called no-definition contains demands from the biomaterials like, for example, no changes in the surrounding tissue and no thrombogenic, allergenic, carcinogenic and toxic reactions [9]. Yet a concept of inertia is questionable as there is no material that does not interact with the body; in the case of "inertia" of a biomaterial there is only a tolerance of the organism [10].

As a result of this insight Williams defined biocompatibility as "the ability of a material to perform with an appropriate host response in a specific application". In Ratner's latest definition biocompatibility even means the body's acceptance of the material, i.e. the ability of an implant surface to interact with cells and liquids of the biological system and to cause exactly the reactions which the analogous body tissue would bring about [2]. This definition requires knowledge of the processes between the biomaterial's surface and the biological system.

Numerous overlapping processes determine biocompatibility. Not only do the mechanical and chemical/physical characteristics of the material influence the tolerance but also the special place of application, the individual reaction of the complement system and the cellular immune system as well as the physical condition of the patient.

The chemical and physical characteristics of the biomaterial's surface which are responsible for the biological reactions at the interface and which, in accordance with Ikada, determine the tolerance of the interface are certainly of great

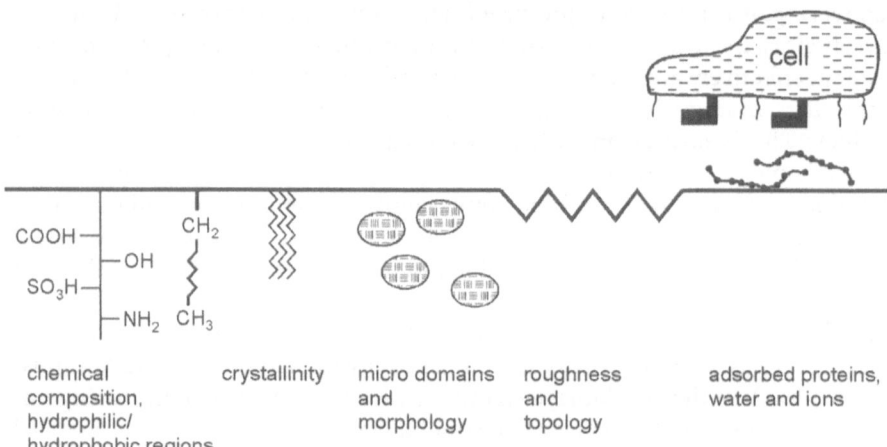

chemical   crystallinity   micro domains   roughness   adsorbed proteins,
composition,   and   and   water and ions
hydrophilic/   morphology   topology
hydrophobic regions

Fig. 2. Interactions of the biomaterial surface and the biosystem

importance [11]. Influencing factors are the chemical structure of the surface,
hydrophilicity, hydrophobicity as well as ionic groups, the morphology, i.e. the
domain structure of a multi-component system such as crystalline and amorphous domains and the topography, i.e. the surface roughness [12]. See Fig. 2
[13].

The surface characteristics concerned can considerably differ from the polymer's bulk characteristics. Due to the minimization of the surface energy and
the chain mobility the non-polar groups move to the phase boundary with air
[14,15]. Additionally, the migration of low molecular components leads to differences between surface and bulk [16,17]. At the phase boundary between the
biomaterial and the aqueous surrounding of the tissue a different situation arises than at the phase boundary between the biomaterial and air. Thus, the surface
characteristics can considerably change after the biomaterial is taken from an
air medium into an aqueous system.

When the implant comes into contact with the biological system the following
reactions are observed:
1. Within the first few seconds proteins from the surrounding body liquids are
   deposited. This protein layer controls further reactions of the cell system. The
   structure of the adsorbed proteins is dependent on the surface characteristics
   of the implanted material. Additionally the adsorbed proteins are subject to
   conformational changes as well as exchange processes with other proteins
   [18].
2. The tissue which borders the implant reacts with dynamic processes which
   are comparable to body reactions in cases of injuries or infections. Due to mechanical and chemical stimuli the implant can lead to a lasting stimulus of inflammation processes. As a consequence a granulated tissue is formed around
   the implant which beside inflammation cells contains collagen fibrils and

blood vessels. A biocompatible implant should thus – as a result of being accepted by the organism – be surrounded with a thin tissue layer which is free of inflammation cells [19].

3. During the course of the contact between the biomaterial and the body the aggressive body medium will cause degradation processes. Hydrolytic and oxidative processes can lead to the loss of mechanical stability and to the release of degradation products [20].
4. As a result of the transport of soluble degradation products through the lymph and vessel system a reaction of the whole body respectively of the concerned organs with regard to the implant cannot be excluded. As well as these processes infection of the biomaterial with bacteria has to be considered as an additional obstacle [21].

The preceding description of the factors which together determine the biocompatibility of an implant shows the diversity of the processes. Until now it has not been possible to completely understand these processes or to comprehend them quantitatively. This understanding is, however, a precondition for the development of biocompatible materials and the prevention of unwanted reactions.

While the term biocompatibility refers to the tolerance of biomaterials with liquid or solid body elements, the term hemocompatibility defines the tolerance of biomaterials with blood. Due to the enormous demand for implants and medical-technical goods for the cardiovascular area, blood tolerance is of great importance. The discussion of blood tolerance, however, demands a separate consideration of the processes between the medium blood and the biomaterial.

1.3
Biomaterial/Blood Interaction

From a clinical point of view, a biomaterial can be considered as blood compatible when its interaction with blood does not provoke either any damage of blood cells or any change in the structure of plasma proteins. Only in this case can it be concluded that this material fulfills the main requests of biocompatibility [9]. As a consequence of the non-specific protein adsorption and adhesion of blood cells, the contact of any biomaterial with blood often leads to different degrees of clot formation [22–24].

The sequence of reactions which take place by the activation of the coagulation system at the blood/biomaterial interface are summarized in Fig. 3. The competitive adsorption behavior of proteins at the biomaterial surface determines the pathway and the extent of intrinsic coagulation and adhesion of platelets. Predictions about the interactions between the biomaterial surface and the adsorbed proteins can only be formulated by having an exact knowledge of the structure of the biomaterial's surface and the conformation of the adsorbed proteins. These interactions are determined both by the hydrophobic/hydrophilic, charged/uncharged, and polar/non-polar parts of the proteins and the nature of the polymer surface [25–27]. A commonly accepted fact is that decreasing sur-

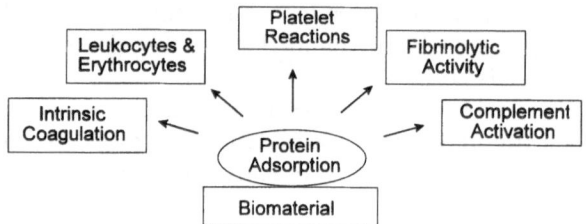

Fig. 3. Blood/biomaterial interactions at the biomaterial interface

face roughness leads to higher compatibility of the material. Reefs which disturb the laminar blood flow, as well as poststenotic turbulence, in particular, can cause clot formation [28].

The role played by the surface tension of a material as one of the most influential factors on protein adsorption is a common subject of discussion. While Andrade defends the opinion that smaller interfacial energies between blood and the polymer surface imply better blood compatibility [29], Bair postulates that a hemocompatible surface should have a surface tension between 20 and 25 mN/m [30]. On the contrary, Ratner provides evidence of good blood compatibility of surfaces with a moderate relationship between their hydrophobic and hydrophilic properties [31]. Others point out the importance of the ionic character of the polymer surface. Biomaterial surfaces with carboxylate, sulfate or sulfonate groups may act as antithrombotic agents as a result of repulsive electrical forces provoked against plasma proteins and platelets [32]. Norde has shown that a decrease in the concentration of ionic groups in the protein and in the polymer surface increases protein adsorption [35]. The relevance of electrical conductivity of biomaterials with respect to blood compatibility is described by Bruck [33]. In addition, the influence of the streaming potential on coagulation has been studied [34,35].

Since fibrinogen activates and albumin inhibits the adhesion of platelets, the adsorption of these proteins has been investigated in many projects. The competitive adsorption of proteins is very complex. In the case of hydrophobic surfaces, fibrinogen is the chief protein adsorbed (not considering hemoglobin), while in the case of hydrogels adsorption of albumin takes place preferably [36,37]. The adsorbed protein film shows time-dependent conformational changes which may cause desorption or protein exchange. Adsorption processes are described by the typical Langmuir isotherms. After long contact times, a stationary state is reached which corresponds to an irreversible protein adsorption [38–40]. The complex time-dependent exchange of proteins is termed the Vroman effect and is observed at every surface with exception of strong hydrophilic ones [41,42]. The quantitative characterization of protein adsorption processes in conjunction with blood coagulation tests as a function of properties of the contact surface is today considered to be an important means for the development of thrombogenic surfaces.

1.4
Biomaterials for Blood Contact: Concepts To Improve Blood Compatibility

By examining the polymeric materials which are currently used in clinical application it can be seen that while their mechanical properties satisfy requirements, their total compatibility with blood has still not been achieved. Therefore, commercial polymers like polyurethanes, silicones, polyolefins and poly(vinyl chloride) which are used as short-term implant materials show thrombogenic properties and require the introduction of anticoagulants [26,43].

Materials which are found in long-term applications count likewise as thrombogenic, e.g. Dacron™ [poly(ethylene terephthalate)], which was introduced some time ago in a woven (soft tissue) or knitted form as a vessel prosthesis with a diameter larger than 6 mm. Here, clot formation is even desired to close the open pores in the textile structure; by fibroplasts and fibrin formation the so-called pseudointima is generated, which prevents blood leaving the endoprosthesis. Due to the strong blood flow and the large diameter of the vessel, closure of the vessel as a consequence of thrombus formation does not occur [44]. This principle is not valid for vessels of diameter less than 6 mm, for the use of these polymers as biosensor membranes, or as artificial organs such as the pancreas.

Under physiological conditions the surface of a natural vessel in contact with blood is built up by a monocellular film of endothelial cells, which carry out regulatory functions of thrombosis, hemostasis and hemodynamic events, and which contribute to the synthesis and transport of compounds which are active in metabolism [45]. The main feature of the endothelial cells is their blood compatibility [46]. As a result of this, the coverage of synthetic surfaces with a mono-

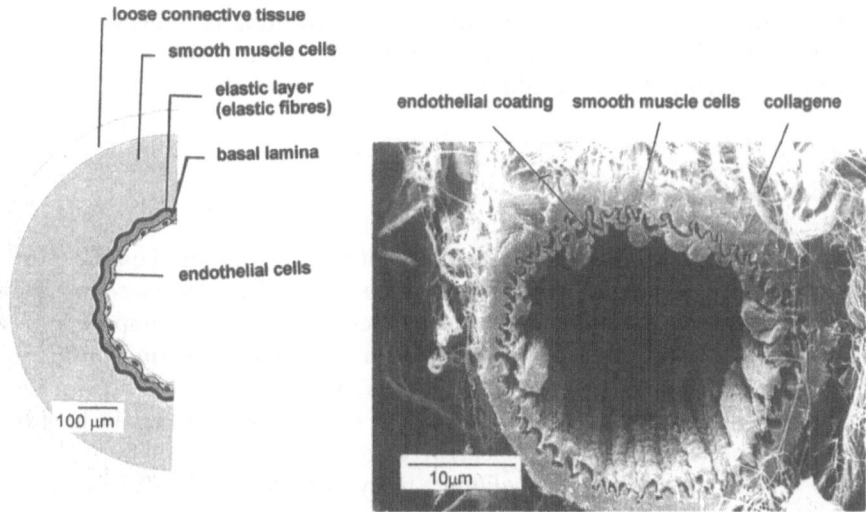

Fig. 4. Structure of a natural blood vessel (a) schematic illustration and (b) scanning electron micrograph

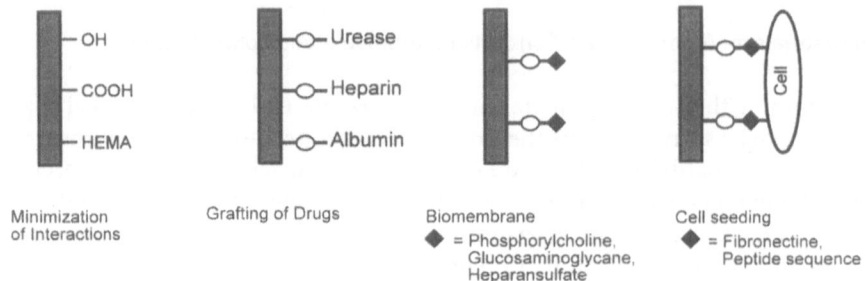

Fig. 5. Concepts to improve the blood compatibility of biomaterial surfaces

layer of human endothelial cells is the most promising concept for obtaining a biocompatible surface, since physiological conditions are closely imitated. The development of such a hybrid organ has not yet been achieved because human endothelial cells do not present any spontaneous growth on foreign surfaces [47].

In a natural vessel the endothelial cells grow on a self-formed basal membrane which is integrated by collagen (types I, III, IV), proteoglycanes and the glycoproteins fibronectin and laminin (Fig. 4). Pierschbacher and Ruoslahti [48–51] discovered that a fragment of fibronectin with the sequence RGD is responsible for the adhesion of endothelial cells. Several research groups have tried to achieve the proliferation of endothelial cells at synthetic polymers by coating their surface with fibronectin or collagen or by covalent coupling of oligopeptides which contain the specific RGD tripeptide sequence [52–55].

The development of an implantation device covered with endothelial cells for introduction as a hybrid organ for long-term application is probably the most complicated and time-consuming task. Further concepts which have been carried out to improve the blood compatibility of biomaterials are given in Fig. 5.

1.4.1
Minimization of Interactions

One way to improve the blood compatibility of polymers is based on the finding that a higher hydrophilicity of the polymer surface implies a decrease in protein adsorption and cell adhesion. In the development of new blood-compatible polymer systems, reduced surface energy has been reached by adjustment of hydrophobic and hydrophilic domains [56,57]. The morphology of the polymer was controlled by an appropriate composition of a polymer blend or by the length of segments in block copolymers. A clearly reduced platelet adhesion was observed on the surface of ABA block copolymers of hydrophilic HEMA (A) and hydrophobic styrene (B) [58]. Special emphasis was placed on polyurethanes, whose chemical structure and surface characteristics were modified with great diversity [56,59,60]. Polyurethanes with alternating hard and soft segments of different hydrophilicity show low fibrinogen and high albumin adsorption [61] which leads to improved blood compatibility [62].

The minimization of biomaterial/biological system interactions is also the goal of defined surface modifications of polymers, without a change in the polymer's bulk properties. For example, upon grafting a polymer surface with PEO the hydrophilicity is increased which provokes a reduction of complement activation and platelet adhesion [37,63]. In the same manner, the introduction of a hydrogel layer, e.g. PHEMA, results in an improvement in the athrombotic properties of the respective hydrophobic polymer [64]. Of special interest is the fixation of functional groups like hydroxyl, carboxyl or amino groups, which do not only reduce the surface energy, but also act as linking groups for further chemical surface modification. For functionalization or monomer grafting, plasma treatment has been proved to be the most suitable surface technique [65].

1.4.2
Grafting of Drugs

The blood compatibility of biomaterials can be improved by means of coating or grafting of anticoagulants, inhibitors of platelet adhesion or fibrinolytically active substances. The most famous example is the ionic or covalent coupling of heparin to the surface of catheters and stents as well as to extracorporal surfaces in contact with blood [66]. Following this principle, oxygenators with covalently bound heparin have been tested in clinical practice. Due to the fact that the activity of the heparin coating decreases with contact time, long-term applications have not yet proved successful. This principle can also be applied to the immobilization of albumin, urokinase or prostaglandin on biomaterials [67].

1.4.3
Imitation of a Biomembrane

A very promising way to avoid any reaction against foreign surfaces consists of the imitation of the membrane of red blood cells as the blood contact surface. In this way Chapman achieved a decrease in platelet adhesion by immobilization of phosphorylcholine at the polymer surface, a main component of the membrane surface of platelets and erythrocytes [68,69]. Other examples following this concept are the coating of biomaterial surfaces with heparin sulfates or glycosaminoglycanes [70]. It is still unknown whether these coatings, which are very active against blood coagulation, are stable enough for longer implantation times. The concepts chosen to improve a biomaterial's compatibility must be selected regarding the planned contact time and the cost.

2
Characterization of Biomaterial Surfaces

The understanding of the interactions which take place between a material surface and the components of the biological system is an important requirement
of biomaterial development. The uppermost atomic layers of a biomaterial,
which present characteristic chemical structural parameters and physical properties, define the contact surface. An important contribution to biomaterial development is, therefore, made by surface-sensitive analytical methods [71]
which allow the surface modifications to the biomaterial to be proved.

2.1
Chemical and Physical Properties of Polymeric Contact Surfaces

2.1.1
Characterization of Chemical Surface Properties

In order to determine the composition and structure of a biomaterial surface
different methods which provide varying degrees of information are commonly
used (Fig. 6). Surface-sensitive infrared spectroscopy supplies the characteristic
absorption bands of functional groups with an informational depth of 0.1–
10 µm by measurement in attenuated total reflectance (IR-ATR). In the case of
samples with rough surfaces photoacoustic spectroscopy (PAS), which allows an
informational depth of approximately 20 µm, can be used [72]. The achieved informational depths are usually larger than the thickness of the modified interface, so that the spectra include information on the bulk composition as well. As
a consequence, surface-sensitive infrared spectroscopy is often not sensitive
enough for the characterization of the modified surfaces.

X-ray photoelectron spectroscopy (XPS) is a more surface-sensitive analytical method which supplies information not only about the type and amount of
elements present but also about their oxidation state and chemical surround-
ings. Applying this method informational depths of approximately 10 nm can be
achieved, which means about 50 atomic layers. In secondary ion mass spectroscopy (SIMS) primary ions interact with the polymer surface and the mass spectra of the formed secondary ions are obtained which give information about the
chemical composition of the outermost atomic layers (approximately 1 nm in
thickness).

Electron spin resonance (ESR) spectroscopy is not applied as a surface-sensitive method. Based on its sensitivity, however, it is possible to detect covalently
bound molecules at the polymer surface if these molecules are labeled with a
spin marker such as 4-amino-TEMPO. The application of atomic force microscopy (AFM) in comparison to scanning electron microscopy (SEM) delivers information about surface properties as far as molecular dimensions. Another advantage of AFM compared with SEM is that the sample is investigated in the
original state (no sputtering) [73]. The full characterization of the surface of a

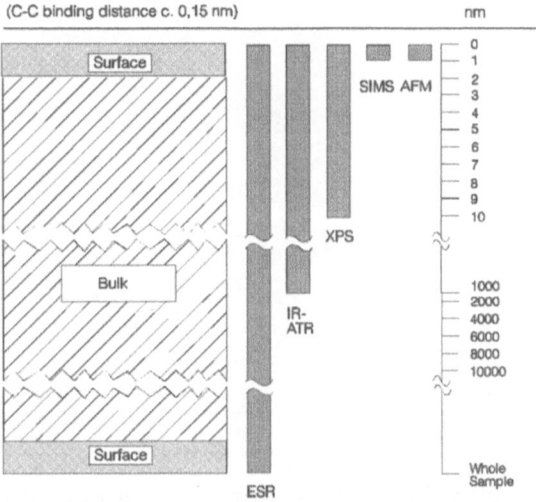

Fig. 6. Informational depths of the spectroscopic and microscopic methods used

biomaterial is often only verified by application of several analytical methods to the sample.

2.1.2
Characterization of Physical Surface Properties

Several methods are available to determine the physical parameters of polymer surfaces. Biomaterials penetrate liquids like blood or water present in soft tissue. It is known that the free surface energy at the biomaterial/water interface is the driving force for the reorientation processes of the polar groups of the uppermost molecular layers of the polymer surface towards the aqueous phase. The chemical composition of the surface of the biomaterial is different depending on its contact with an aqueous medium or with air. Hydrophilic domains of polymer systems like those found in block copolymers, for example, are mostly located at the aqueous interface, while the hydrophobic ones tend to remain at the air interface. The investigation of surface wettability by means of contact angle determination and the measurement of the streaming potential (ζ-potential) is of special interest in the characterization of the polymer surface.

The contact angles of water and suitable solvents at the solid/liquid/gas interface allow the determination of the surface tension of solids as well as the dispersive (γ^d) and polar (γ^p) components. A semiquantitative prediction of the hydrophilicity and hydrophobicity of polymer surfaces has already been achieved by contact angle determination with water [74,75].

The contact angle determination at surfaces of biomaterials is carried out by several methods which are schematically presented in Fig. 7, i.e.

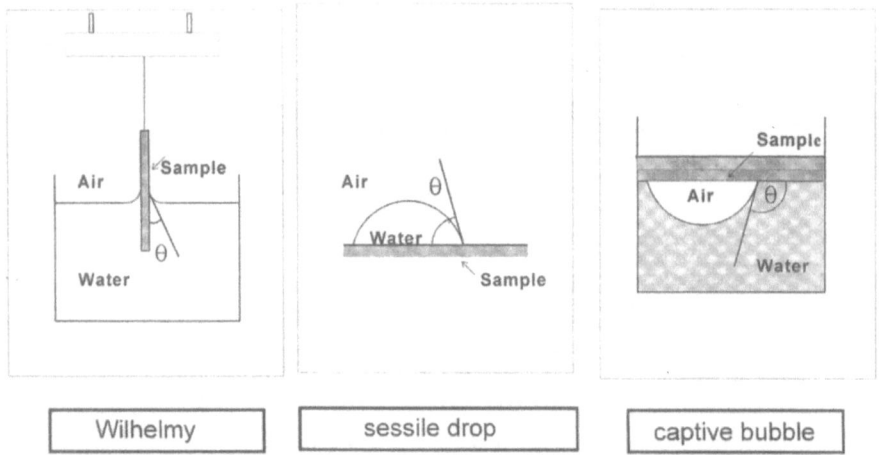

Fig. 7. Schematic illustration of dynamic and static contact angle determinations using the Wihelmy plate method, the sessile drop method and the captive bubble method

- the Wilhelmy plate method [76,77]
- the sessile drop method [78]
- the captive bubble method [79]

A distinction can be made among the available methods between static and dynamic contact angle determination methods. In the case of a static determination the contact angle of a drop with an immobile solid/liquid/gas interface is determined microscopically (sessile drop). In the captive bubble method the contact angle of an air bubble, which is located under the solid surface in contact with the liquid, is determined. In contrast to the sessile drop method, in the captive bubble method the contact angle is measured at a completely wet surface.

The Wilhelmy plate method is a dynamic method which permits the contact angle determination between a sample with perimeter p and a liquid with a surface tension γ_l during immersion and withdrawal of the sample [80] measuring the force F. The contact angles on immersion θ_a (advancing angle) and on withdrawal θ_r (receding angle) are calculated according to Eq. (1):

$$F_{a,r} = \gamma_l\, p \cos \theta_{a,r} \tag{1}$$

The difference between the advancing angle θ_a and the receding angle θ_r is called the contact angle hysteresis.

Electrochemical properties are other important physical surface parameters. The existing surface charge density, i.e. the surface potential, has a strong influence on protein adsorption and blood compatibility [81]. In this way the characterization of the interface charge density of a biomaterial by ζ-potential determination delivers an important parameter for understanding blood compatibility of biomaterial surfaces [82–85].

2.2
Biological Parameters To Describe Biological Interactions

In vitro assays are nowadays applied to characterize the biocompatibility of newly developed polymer systems and surface-modified polymers. The biological parameters obtained give valuable information about a possible incompatibility and decrease the number of animal studies. The following in vitro experiments have found application:
- Enzyme-linked immunosorbent assay (ELISA) to determine protein adsorption.
- Cell culture approaches to investigate *in vitro* cell growth behavior.
- Blood compatibility investigations.

2.2.1
Enzyme Linked Immunosorbent Assay (ELISA) To Determine Adsorbed and Immobilized Proteins

To detect the adsorption of a protein at a biomaterial's surface appropriate ELISA tests, based on the specific reaction of antibodies with surface-bound antigens, are applied [86]. An indirect ELISA variant in which a primary antibody binds to the adsorbed protein (antigen) is shown in Fig. 8. In a second step a secondary antibody, which is conjugated with an enzyme like horse radish peroxi-

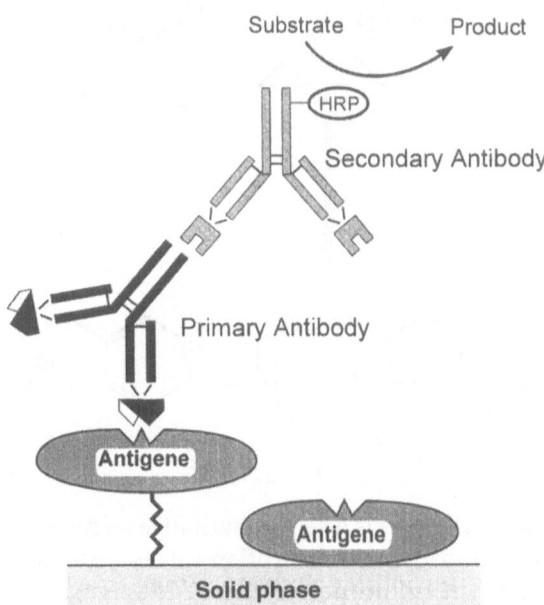

Fig. 8. Schematic illustration of the reaction steps of the determination of the protein adsorption and immobilization using the enzyme-linked immunosorbent assay

dase, binds to the primary antibody [87]. Then this antigen-antibody-antibody complex reacts with a substrate solution. In the presence of hydrogen peroxide and by means of the enzyme the leuco form of a dye is oxidized and color is developed, the optical density being proportional to the amount of bound antigen. The substrate is diammonium 2,2'-azinobis(3-ethylbenzothiazoline sulfonate) (ABTS), whose oxidation product is a metastable, green radical cation with absorption maxima at 410 and 650 nm, which finally undergoes disproportionation and leads to the dication and ABTS (Eq. (2)) [88].

The described assay (ELISA) is a semiquantitative technique and is based on a comparison of the spectroscopic absorption values with those obtained with reference materials. Tissue culture polystyrene (TCPS) is used as a reference surface with a high protein-binding capacity. The detection of the binding capacity of antibodies is useful to establish the amount of adsorbed and immobilized fibronectin. When a high amount of adsorbed or immobilized fibronectin exists at the polymer surface, enhanced growth of the cells, e.g. of endothelial cells, can be expected. With respect to blood compatibility of biomaterial surfaces, the adsorption behavior of the plasma protein fibrinogen is important [89]. Low fibrinogen adsorption may indicate good blood compatibility.

2.2.2
In Vitro Cell Growth

The highly sensitive system of a cell culture, which permits the determination of cell growth influencing factors and thereby the selection of cell toxic materials, is used to judge the biocompatibility of biomaterials. In addition, cell culture experiments allow the investigation of tissue compatibility of biomaterials and help to reduce the number of animal studies [90]. In cell culture investigations two cell types can be used: organ specific cells and cells belonging to permanent cell lines which have lost their specificity. Organ specific cells have the specific properties of the tissue from which they were obtained. Primary culture cells consist of original tissue cells which are being used in the cell culture for the first time and which can be utilized to obtain secondary cultures [91]. In this way the human umbilical vein endothelial cells (HUVEC) used for cell seeding experiments are only cultivated up to the second passage (Fig. 9).

In order to determine whether a biomaterial's surface is suitable for a later cell seeding, proliferation studies with HUVEC are carried out over several days.

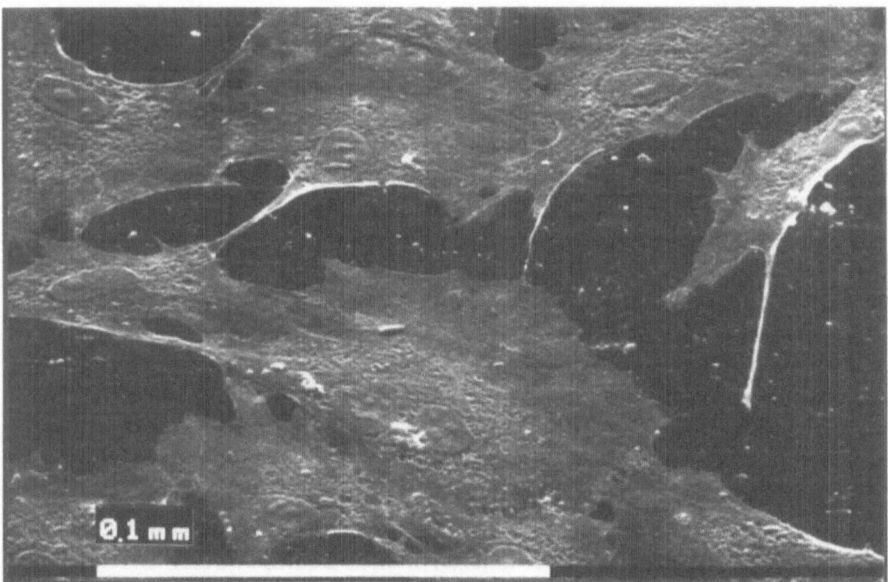

Fig. 9. Micrograph of a polymer surface with human umbilical vein endothelial cells (HUVEC)

The evaluation is done by measurement of the cell density after the first, fourth and seventh cultivation day at the samples's surface and by comparison of these values with those obtained in the positive control Thermanox [92].

2.2.3
Blood Compatibility Investigations

The contact of biomaterial surfaces with the biological system blood provokes, with different intensity, activation of the intrinsic coagulation pathway at the blood/biomaterial interface. Clinically important and reproducible investigation methods are carried out to evaluate blood compatibility. The following coagulation parameters, obtained after the contact of the foreign surface with native, non-anticoagulant human whole blood in a modified Bowry blood chamber [93] and compared to the initial citrate plasma values, are evaluated:
- Activated partial thromboplastin time (aPTT),
- Partial thromboplastin time (PTT),
- Platelet adhesion/platelet number, and
- Leucocyte number

Both aPTT and PTT measure the coagulation factors of the intrinsic system during a defined activation time until clotting occurs. The difference between both tests resides in the fact that in aPTT kaolin is added to provoke an additional contact activation of factor XII.

The most important parameters to characterize blood compatibility are thrombocyte adhesion and thrombocyte number. In contact with a foreign body platelets tend to adhere similarly as in the case of an external injury. For this reason materials which show strong platelet adhesion as a consequence of their contact with a foreign body or provoke a decrease in the number of blood platelets are considered as thrombogenic [94]. The decrease in the quantity of blood leucocytes after contact with a foreign body is a sign of a "cellular immunoresponse" of the biological system toward the biomaterial.

2.3
Correlation of Physical Surface Properties with Blood Responses

A number of polymer surface parameters, e.g. surface free energy and surface charge, are responsible for blood interaction phenomena. Regarding the correlation of hydrophilicity and thrombocyte adhesion, Ikada et al. determined a maximum of thrombocyte adhesion for a contact angle region between 60 and 80° [95]. Van Wachem et al. showed that the best blood compatibility of polymer blends is achieved for moderate wettability [96]. The influence of polar and dispersive components of the surface tension on blood compatibility was described by Kaelble and Coleman [97,98]. They found that polymers with high dispersive (γ^d) and low polar components (γ^p) of surface tension show better blood compatibility than polymers with low dispersive interactions. Furthermore, a nega-

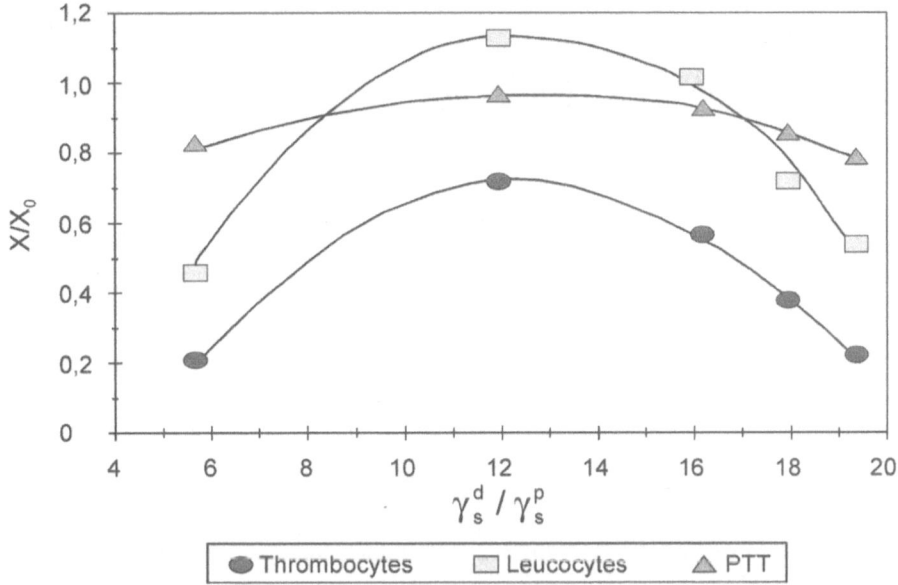

Fig. 10. Blood compatibility parameter X/X_0 as a function of γ_s^d/γ_s^p after the blood interaction with the water-extruded PPE/EVA blend (X_0: value without surface interaction)

tively charged polymer surface prevents blood coagulation because of electrostatic repulsion between the polymer and the platelets or plasma proteins [99]. In order to transmit these results to the *in vivo* situation it has to be mentioned that the endothelial cell layer of the inner wall of a natural vessel shows a slightly negative ζ-potential (between –8 and –13 mV) [34].

Water-extruded blends of poly(ethylene-*co*-propene) with various amounts of poly(ethylene-*co*-vinyl acetate) (PPE/EVA blends) were prepared and the wettability and the ζ-potential were measured [100]. The relationship of the varying wettability and surface charge values and defined blood compatibility parameters is shown in Figs. 10 and 11. Figure 10 demonstrates the blood compatibility parameters (APTT, number of thrombocytes and leucocytes) after blood/material interaction with respect to the parameters before exposition X/X_0 as a function of the ratio of the dispersive and the polar component of the surface tension (γ_s^d/γ_s^p). The best blood compatibility, that means $X/X_0=1$, was obtained for a ratio of the dispersive and polar component of ca. 12, which corresponds to the surface tension of the PPE/EVA blend with an EVA content of 50 wt%. Probably the balanced ratio between the hydrophilicity and the hydrophobicity of the PPE/EVA blend (50:50) is responsible for the good blood compatibility.

Figure 11 shows X/X_0 as a function of the ζ-potential. The ζ-potential was determined in a 10^{-3} M KCl solution at the physiological pH value of 7.2. Comparable parabolic curves were obtained with a maximum of blood compatibility within a ζ-potential range between –4 and –8 mV. The ζ-potential of the water-

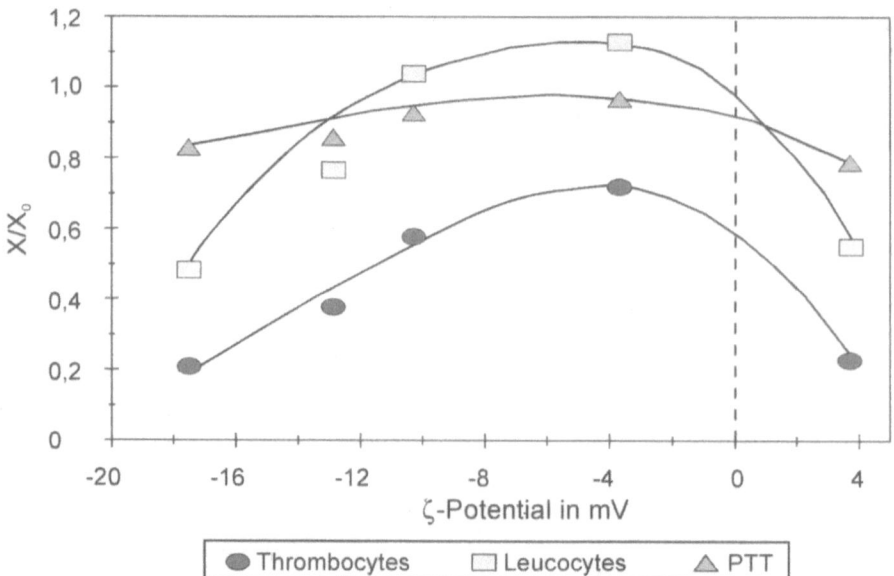

Fig. 11. Blood compatibility parameter X/X_0 as a function of the ζ-potential (pH 7.2) after the blood interaction with the water-extruded PPE/EVA blends (X_0: value without surface interaction)

extruded PPE/EVA blend (50:50) amounts to –4 mV. Thus the results of surface tension and ζ-potential coincide.

The existence of these optimum values may be due to a biocompatibility window. However, this only applies to PPE/EVA blends. It has to be expected that the biocompatibility window of other polymer systems is displaced due to further influences of material-related surface properties.

3
Possibilities of Improving the Interfacial Biocompatibility of Polymers

Depending on the kind of application either an improved tissue compatibility in soft tissue use or an improved blood compatibility in the case of polymer surfaces in contact with blood are sought. An appropriate surface of a polymer system can be adjusted by selecting a defined ratio of hydrophilic and hydrophobic groups. The modification of a given polymer surface is of special interest.

The objectives may be:

- Hydrophilization to "inactivate" the biomaterial's surface,
- Functionalization of the polymer surface to increase fibronectin adsorption, or
- Immobilization of cell adhesion mediators on polymer surfaces to induce permanent cell seeding

Low-temperature plasma treatment is an important means to achieve these objectives without alteration of the bulk properties and without the use of solvents or reagents to be removed later in the process [101,102]. Plasma "etching" with non-polymerizable gases like argon, nitrogen or sulfur dioxide, was introduced to hydrophilization and functionalization [103]. The plasma-induced graftcopolymerization of monomers with suitable functional groups can be applied for subsequent immobilization reactions.

3.1
Inactivation of Polymer Surfaces

The minimization of chemical and physical interactions between polymer and blood plasma is presently the actual dominant concept to improve the blood compatibility of polymers for intra- and extracorporal short-term applications [104,105]. Most polymers for tube and catheter systems present a hydrophobic surface. Plasma technology is applied to hydrophilization and graftcopolymerization of hydrophilic monomers [106,107].

This approach has been used to hydrophilize the surface of the cycloaliphatic polyurethane Tecoflex™ which is the material used for catheter systems. Upon plasma treatment radicals are formed on the polymer surface which in contact with air react with oxygen and form hydroperoxides [108]. The subsequent photoinitiated or thermal decomposition of the hydroperoxides produces secondary radicals that are able to initiate the polymerization of suitable monomers [109] (Fig.12). 2-Hydroxyethylmethacrylate (HEMA), hydroxybutylacrylate (HBA) or acrylic acid (AAc) can be used as hydrogel building monomers.

Hydrogel layers are known to provoke low accumulation of blood proteins and low reaction with cellular blood components like platelets in comparison to hydrophobic surfaces, due to their strong affinity to water and their low surface energy [110].

The hydrophilization of poly(ether sulfone) surfaces used for the production of ultrafiltration membranes for hemodialysis is of special interest. Poly(ether sulfone) is a chemically inert and highly thermostable polymer showing a hydrophobic surface which leads to high fibrinogen adsorption and subsequent thromboembolization [111]. This effect is generally avoided by non-desirable heparin doses. The hydrophilization of poly(ether sulfone) surfaces by plasma-induced graftcopolymerization and the introduction of a hydrogel layer without

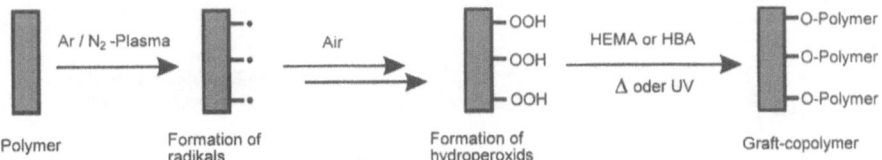

Fig. 12. Schematic illustration of plasma-induced graftcopolymerization of hydrogels on polymer surfaces

affecting the bulk properties seems to be a suitable approach to increase blood compatibility.

3.1.1
Hydrophilization of Tecoflex™ by Plasma Treatment

Special efforts are being carried out to improve the blood compatibility of bio-materials which are already in use for medical devices. In order to hydrophilize the surface of the commonly used cycloaliphatic poly(ether urethane), Tecoflex and argon (Ar) and sulfur dioxide (SO_2) plasma treatment has been applied. The surfaces of treated Tecoflex foils were characterized as a function of plasma treatment time (10, 60 and 300 s). XPS shows SO_x groups in the Tecoflex surface after SO_2 plasma treatment. The results of the dynamic contact angle determination of Ar or SO_2 plasma treated Tecoflex surfaces are presented in Fig. 13.

A significant increase in the wettability of the Ar or SO_2 plasma treated Tecoflex surfaces with increasing plasma treatment time is observed. The effect is particularly evident in the receding angle, which is more sensitive to changes in the hydrophilicity of the surface than the advancing angle. Under air storage, as a consequence of the minimization of the interfacial energy, a migration of polar groups into the inside of the polymeric material is observed and hence a decrease in hydrophilicity. A renewed increase in hydrophilicity, i.e. a decrease in the contact angle of the investigated surfaces, is achieved by storage in water for one week due to reorientation processes.

The fibrinogen adsorption determined by means of the ELISA technique gives first evidence of the enhanced blood compatibility of the modified Tecoflex

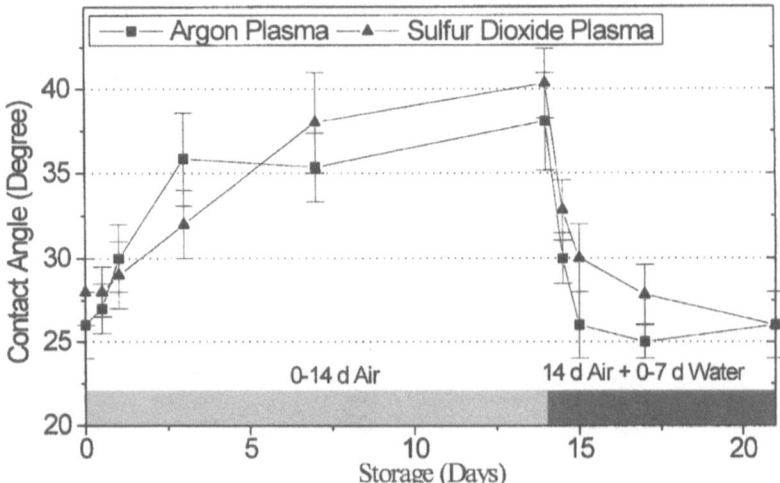

Fig. 13. Contact angle of Tecoflex after 30 s SO_2 and Ar plasma treatment as a function of time and environment

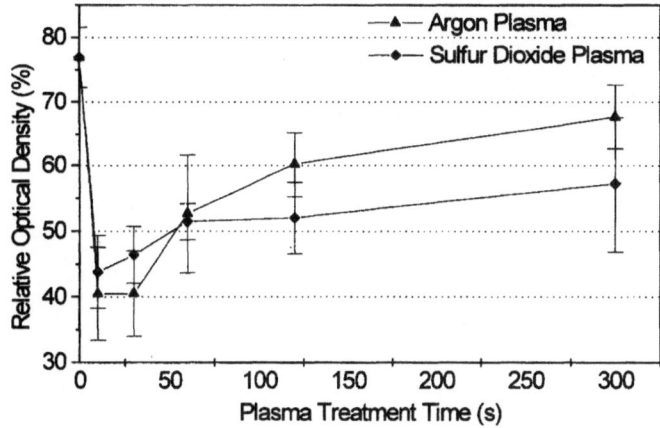

Fig. 14. Fibrinogen adsorption on Ar and SO$_2$ plasma treated Tecoflex™ as a function of plasma treatment time, measured by ELISA, data relative to TCPS (100%)

surfaces (Fig. 14). Fibrinogen adsorption is reduced by 50% when increasing the hydrophilicity of the polymer surface by Ar or SO$_2$ plasma treatment for short times (10 and 60 s). The increase in fibrinogen adsorption for the 300 s plasma-treated surface is attributed to the etching effect and the creation of surface roughness.

When the fibrinogen adsorption is reduced upon hydrophilization of Tecoflex surfaces, an improved blood compatibility is expected. Since the hydrophilicity of the surface under common storage conditions is not permanent the development of stable hydrophilic polymer surfaces is desirable.

3.1.2
Plasma-Induced Graftpolymerization of Hydrogels on Poly(ether sulfone) and on Tecoflex™

Plasma-induced grafting of hydrogels offers a possibility to obtain permanent minimization of the surface energy of hydrophobic polymer surfaces. Hydroxyethylmethacrylate (HEMA) was applied as hydrophilic monomer, the polymer of which is already known in the fabrication of contact lenses [1,112]. Acrylic acid (AAc) was chosen as the hydrogel building monomer to generate a stable, hydrophilic surface with the additional possibility of using the acid functions for further immobilization of bioactive substances.

Hydrogel Coating of Poly(ether sulfone). Due to the fact that PES has no functional groups which allow a wet chemical modification, plasma-induced hydroperoxide formation presents a suitable method to carry out the subsequent thermoinitiated graftpolymerization of HEMA (cf. Fig. 12). The iodometric determination of the hydroperoxide concentration at the surface shows that a saturation value is achieved after 4 min of argon plasma treatment and subsequent air exposure [113].

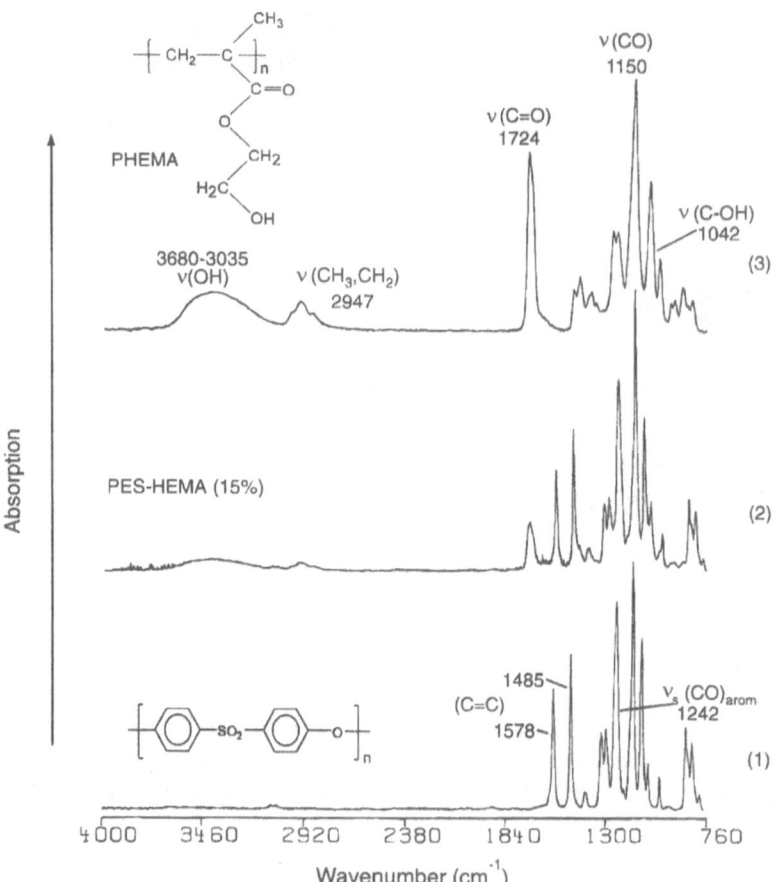

Fig. 15. IR-ATR spectra of unmodified PES (*1*), HEMA-grafted PES (15 wt% monomer so-lution) (*2*) and PHEMA (*3*) (Ge-crystal with an informational depth of 0.1–1 μm)

The HEMA grafting was proved by means of spectroscopic methods with dif-ferent informational depths. The elemental composition of the PES surface grafted with HEMA determined by XPS shows a content of sulfur of 0.6%. This value falls within the detection limits of XPS and points to the fact that an ap-proximately 10 nm thick hydrogel layer of PHEMA is formed on the PES surface.

The IR-ATR spectra of the untreated and modified PES as well as that of the PHEMA reference surface are shown in Fig. 15. The spectra of the modified PES surface shows the superposition of the bands of the pure PES and PHEMA. The surface tension of the PES surface modified with the hydrogel was investigated by means of the captive bubble technique by measurement of the air/water and octane/water contact angle (Fig. 16). The polar and dispersive components of the surface tension, γ^p and γ^d, were calculated from the contact angles deter-mined [114].

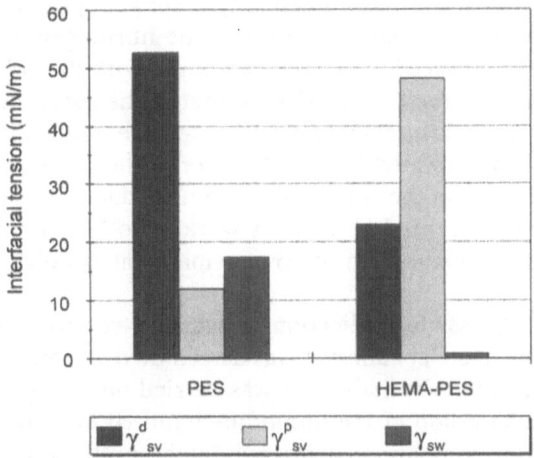

Fig. 16. Interfacial tension γ_{sw} (whole and polar γ^P_{sv} and dispersive γ^d_{sv} portions) of the HEMA-grafted PES surface (HEMA-PES) in comparison with the unmodified PES surface

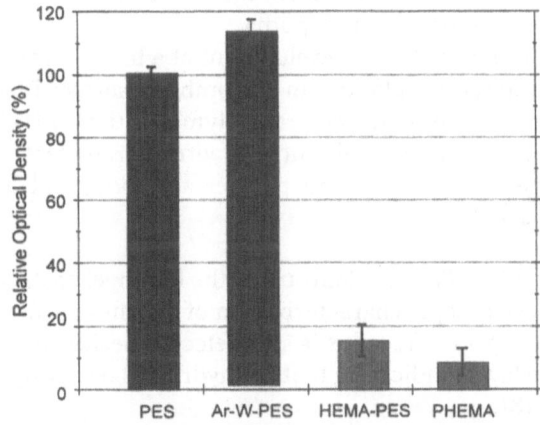

Fig. 17. Fibronectin adsorption of the unmodified and modified PES surface (HEMA-PES) in comparison to the Ar plasma treated PES surface (Ar-W-PES) and the PHEMA reference surface determined by ELISA

The surface tension at the solid/air interface γ_{sv} and at the solid/water interface γ_{sw} is the sum of the polar and the dispersive components [115]. In comparison with the untreated PES surface which has a surface tension at the solid/water interface γ_{sw} between the solid and the physiological relevant medium water of 17 mN/m, the HEMA-grafted surface shows a strong decrease in surface tension γ_{sw} to approximately 0 mN/m.

The fibrinogen adsorption at the HEMA-grafted PES surface was studied by means of the ELISA technique in comparison to that of pure PHEMA. The fibrinogen adsorption at the unmodified poly(ether sulfone) surface was used as the standard (Fig. 17). A clear reduction in the fibrinogen adsorption at the HEMA-grafted PES surface is observed compared with the poly(ether sulfone) surface. The value observed is identical to that at the surface of pure PHEMA within the limits of experimental error. The increase in the fibrinogen adsorption at the Ar plasma treated PES surface contradicts the usual observation [116]. A reason might be the surface enlargement due to plasma etching. The present effect was confirmed by Feijen's work, who likewise observed an increase in fibrinogen adsorption at Ar plasma treated polyethylene surfaces [117].

Considering the possible application of hydrogel-coated poly(ether sulfone) in blood contact the platelet adhesion was determined as a biological parameter. After incubation, qualitative judgment was carried out using atomic force microscopy. Platelet adhesion of the unmodified and HEMA-grafted PES surface was investigated in comparison with poly(ether urethane) (Pelletane 2363–90AE) as a negative control (Fig. 18).

As a consequence of the hydrophobic character of PUR and the unmodified PES surface, platelets have a spread-out form. The hydrogel surface is, on the contrary, almost free of platelets which, in the case of adhesion, exerts a spherical form without formation of pseudopodia.

In summary, with regard to the development of a blood-compatible polymer surface based on poly(ether sulfone), an athrombotic surface is achieved by hydrogel coating using plasma-induced graftpolymerization of HEMA. Nevertheless, this evaluation is based on only three in vitro parameters, i.e. blood compatibility, fibrinogen adsorption, and platelet adhesion, and hence is by no means comprehensive.

Hydrogel Coating of Tecoflex™.   Contrary to the hydrogel coating of poly(ether sulfone), in the spectroscopic characterization of Tecoflex surfaces grafted with HEMA and AAc using XPS, Tecoflex is still detected beside the hydrogel in the uppermost surface layers indicating that the hydrogel layer is significantly thinner than 10 nm [118].

In view of further immobilization of bioactive substances via the carboxylic functions of the grafted acrylic acid, investigation of the accessibility of the acidic groups was carried out by means of appropriate derivatization reactions with TFE using the coupling reagent EDC (Eq. (3)) and detection of the reporter element fluorine by means of XPS.

$$\text{[structure]} \quad (3)$$

The surface composition obtained using XPS shows that Tecoflex™ grafted with AAc and reacted with TFE contains 9.8 atom% fluorine, while in the surface of the original polymer as a reference no fluorine is detected. The CF_3 line at a

Negative control
PUR

PES

HEMA-PES

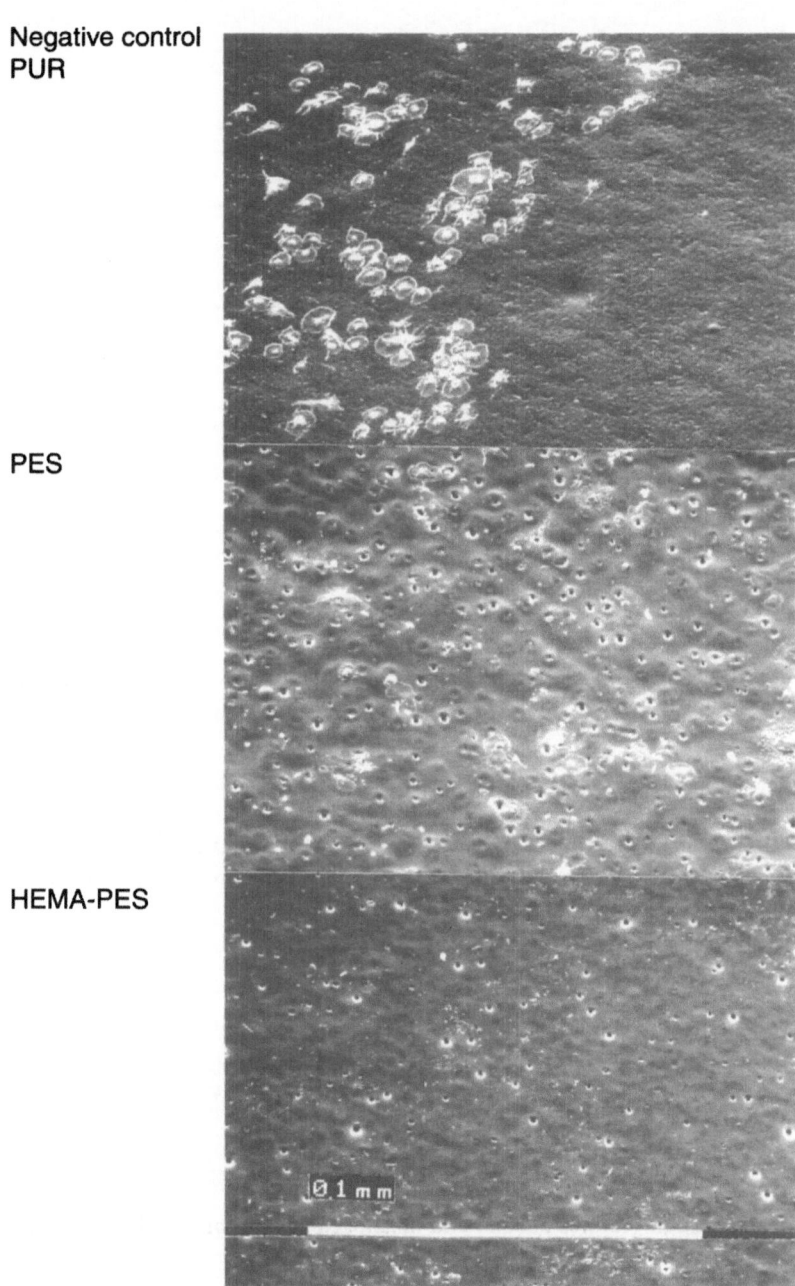

Fig. 18. SEM micrographs of the thrombocyte adhesion on poly(ether sulfone) (PES) HEMA-grafted PES (HEMA-PES) and the negative control polyetherurethane (PUR), magnification ×600

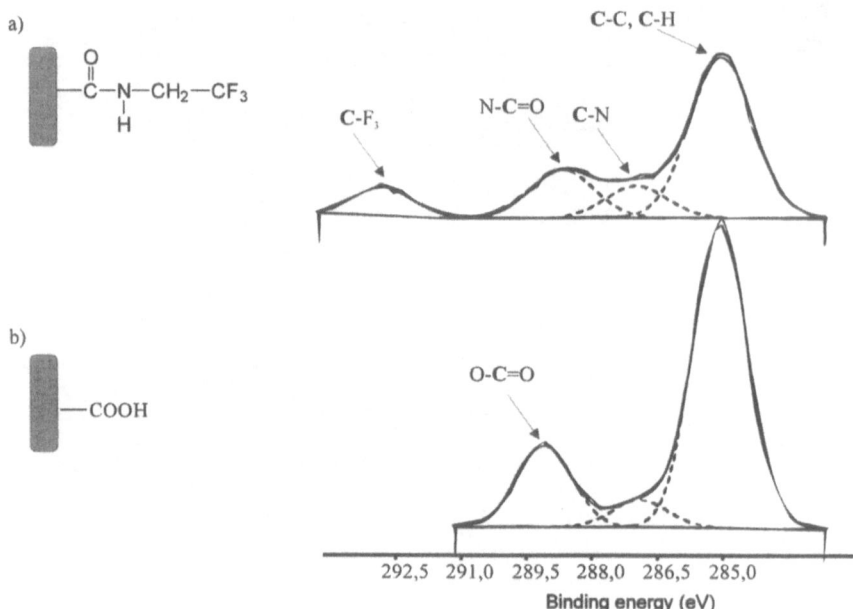

Fig. 19. Spectra from C_{1s} XPS of AAc-grafted Tecoflex before (*a*) and after (*b*) derivatization with 2,2,2-trifluoroethylamine

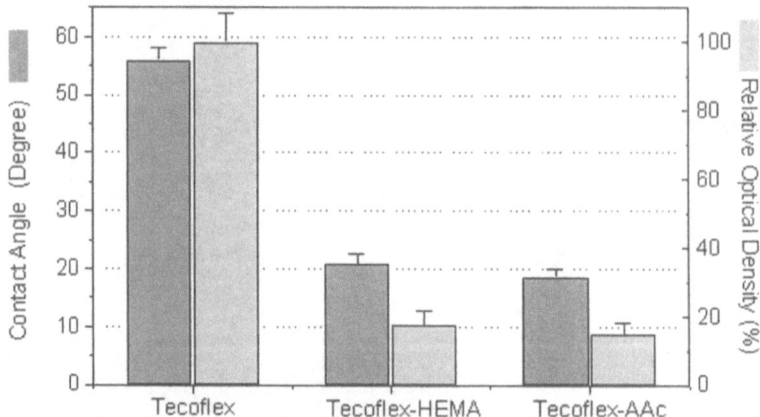

Fig. 20. Contact angle (captive bubble) and fibrinogen adsorption (ELISA) of HEMA and AAc-grafted Tecoflex in comparison with untreated Tecoflex

binding energy of 292.9 eV is characteristic of the high resolution C_{1s} element spectrum of the surface-modified Tecoflex™ (Fig. 19).

The determination of the contact angle was carried out by means of the captive bubble method. The influence of the wettability on fibrinogen adsorption determined by means of the ELISA method at the hydrogel-coated Tecoflex surface, compared with the uncoated surface, is presented in Fig. 20. Hydrogel coating of Tecoflex leads to a strong reduction in fibrinogen adsorption with a simultaneous increase in hydrophilicity in comparison with the unmodified surface. No significant difference between PHEMA and PAAc coatings is observed.

3.2
Functionalization of Polymer Surfaces To Increase Fibronectin Adhesion

The intimate contact between implant and tissue is a decisive prerequisite for the biocompatibility of long-term implants in soft tissue application. To achieve this an interface must be created which provokes optimal cell growth. As for the

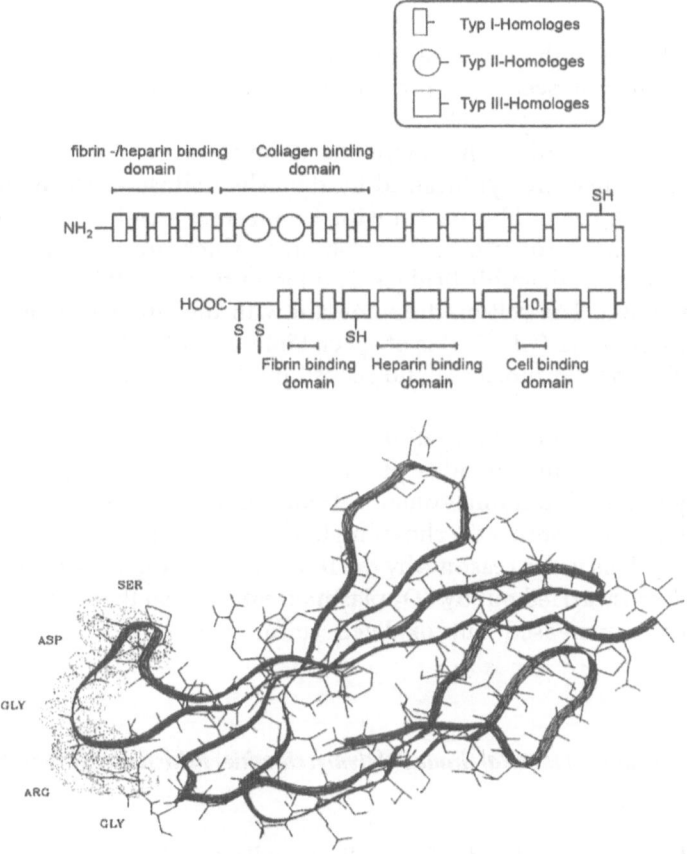

Fig. 21. Model of a subunit of the fibronectin molecule according to MC-Donagh [85] and the cell binding domain according to Dickinson and Ruoshlahti et al. [86]

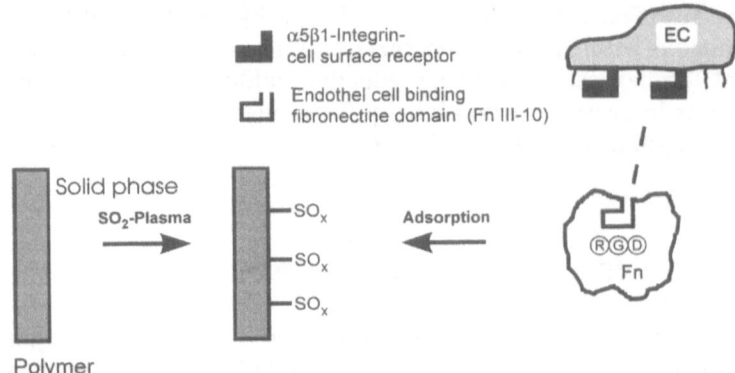

Fig. 22. Functionalization of polymer surfaces to increase the bibronectin adsorption using SO_2 plasma [11]

natural structure of the extracellular matrix (ECM), the isolated extracellular protein fibronectin seems to present a suitable interface for cell growth [54,119,120].

Fibronectin is an adhesion protein like laminin, vitronectin, and von Willebrand factor, which are synthesized by the cells themselves to build up the ECM. The glycoprotein fibronectin with a molecular weight between 220,000 and 250,000 consists of two similar subunits, which are connected close to their C-terminus by disulfide bridges. The subunits are composed of functional domains [121]. The cell binding domain with the characteristic sequence Gly-Arg-Gly-Asp-Ser (GRGDS) is of special interest [122]. Models of the subunit of the fibronectin molecule and its cell binding domain are presented in Fig. 21.

The objective of functionalization of polymer surfaces is to create a surface that shows high fibronectin adsorption and, at the same time, guarantees the availability of its cell binding domains. Hydrophilic surfaces favor adsorption of fibronectin and fibronectin shows high affinity to sulfonic acid and sulfate groups [123]. This is the reason why the functionalization of hydrophobic polymer surfaces is carried out by SO_2 plasma treatment, so that not only the hydrophilicity is increased but oxidized sulfur groups are also introduced (Fg. 22).

3.2.1
Surface Modification of Medical Grade Poly(vinyl chloride) to Increase Fibronectin Adsorption

Plasticized poly(vinyl chloride) for medical application is a copolymer of ethylene, vinyl acetate and carbon monoxide which is grafted with vinyl chloride (PVC/EVACO) and does not contain any low molecular weight plasticizer. Upon saponification of the vinyl acetate groups of the PVC/EVACO surface hydroxy groups suitable for surface modification are created (Eq. (4)).

$$\text{(4)}$$

The saponification step is accompanied by the disappearance of the photoline of the C=O carbon of the acetate group at a binding energy of 282.2 eV in the C_{1s} XPS spectrum. In addition, in the IR-ATR spectrum a broad OH stretching vibration band appears at 3410 cm^{-1} (Fig. 23). According to Clark and Briggs, on subsequent SO_2 plasma treatment, sulfate groups are formed [124,125]. SO_x functions may be incorporated into the poly(vinyl chloride) sequences as well.

Spectra from XPS, especially the S_{2p} spectra, confirm the incorporation of SO_x groups after SO_2 plasma treatment of the saponified PVC/EVACO films (Fig. 24). The oxygen content of the surface is increased to 17.8 and the sulfur content to 6.8 atom% with the $S_{2p}3/2$ photoline being located at 169.6 and 168.8 eV due to sulfonate and sulfate groups, respectively [126,127].

In order to functionalize plasticized PVC, the saponified polymer surfaces were reacted with hexamethylene diisocyanate (HDI) as a spacer and subsequently with ethylene glycol, 1,3-propanediol and 1,4-butanediol (Eq. (5)).

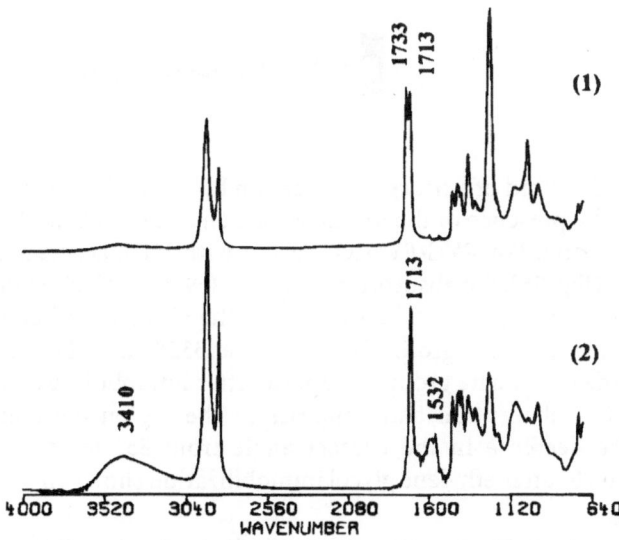

Fig. 23. IR-ATR spectra of unmodified (1) and saponified (2) PVC/EVACO surface (Ge-crystal, informational depth: 0.1–1 µm)

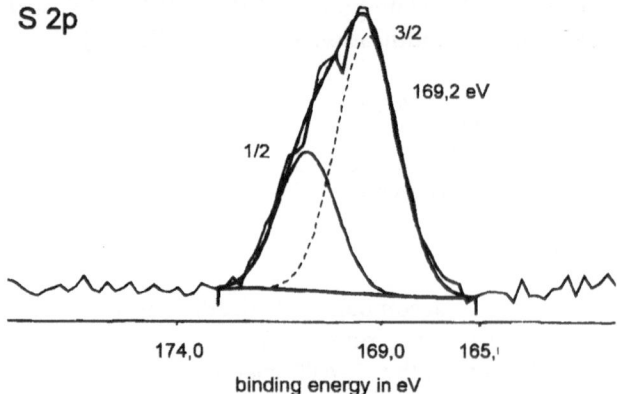

Fig. 24. Spectrum from S_{2p} XPS of the saponified SO_2 plasma treated PVC/EVACO film (treatment time: 6 min)

Thus, microdomains rich in hydroxyl groups are introduced via urethane bonds which exert good hydrolytical stability [128].

(5)

XPS proves this kind of surface modification by indicating the nitrogen content as well as the presence of the urethane carbon at 289.3 eV in the high-resolution C_{1s} spectrum. For PVC/EVACO reacted with HDI, according to the IR-ATR spectrum (Fig. 25), the absorption band at 2269 cm^{-1} gives evidence for the terminal isocyanate groups. At the same time, the absorption band due to the NH groups of the urethane groups is observed at 3328 cm^{-1}. The characteristic bands due to the isocyanate groups disappear after immobilization of the diols, and a broad OH absorption band appears in the region between 3410 and 3330 cm^{-1}. The decrease in the contact angle from 85° for the unmodified PVC/EVACO to 70° after ethylene glycol immobilization shows an increase in the hydrophilicity.

The adsorption of fibronectin on unmodified and modified PVC/EVACO films as determined by means of a suitable ELISA is presented in Fig. 26 [129]. The modified PVC/EVACO surface with ethylene glycol attached to it shows an

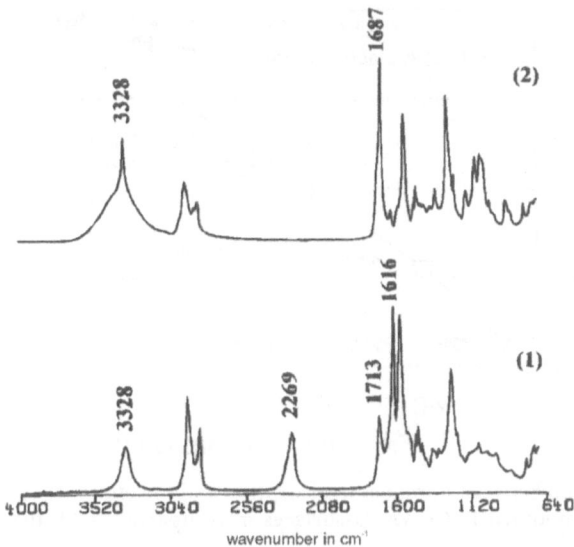

Fig. 25. IR-ATR spectra of the HDI-activated (*1*) and ethylene glycol immobilized (*2*) PVC/EVACO surface

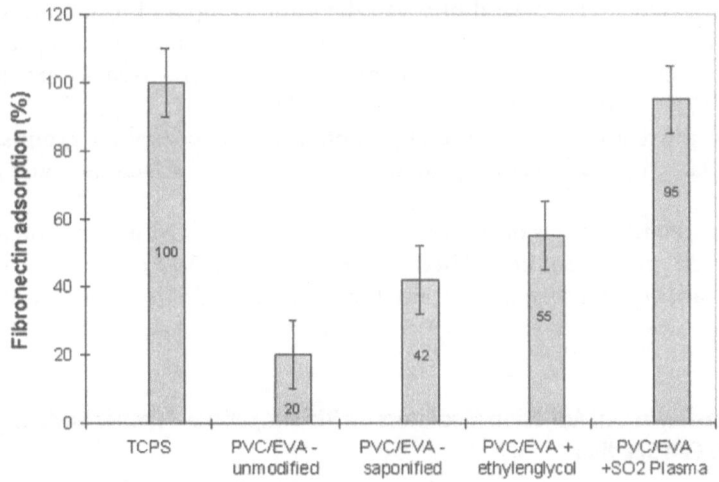

Fig. 26. Fibronectin adsorption of unmodified and modified PVC/EVACO films in comparison with the positive control surface (TCPS)

increase in fibronectin adsorption of 20–30% in comparison with the unmodified surface. It is worth emphasizing that fibronectin adsorption at the SO_2 plasma treated PVC/EVACO surface achieves 95% with respect to the reference surface (100%).

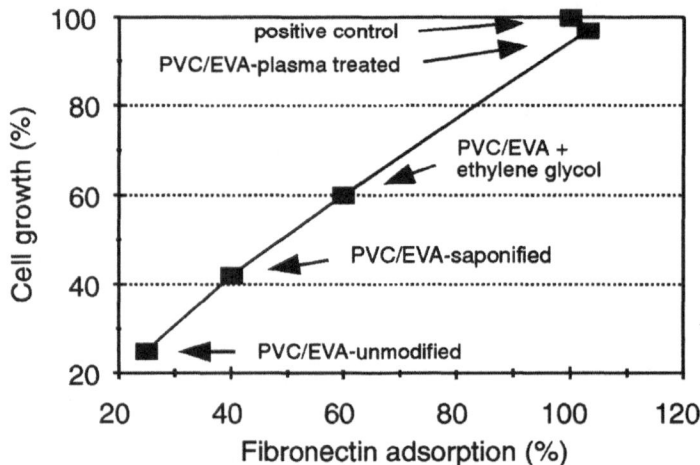

Fig. 27. Relationship between fibronectin adsorption and the endothelial cell proliferation modified and unmodified PVC/EVACO surfaces in comparison with the control surface Thermanox

The growth of endothelial cells (HUVEC) at modified PVC/EVACO surfaces was investigated over a period of seven days and compared to the unmodified surface and to the positive standard Thermanox [130]. After seven days the highest proliferation rate was found at the SO_2 plasma treated surface and achieved 103% of that of the positive control. The contact angle of 67° is a sign of a moderate relationship between hydrophobic and hydrophilic groups, which was postulated by Ikada as the optimal condition for cell attachment and growth [131].

A linear relationship is observed between the fibronectin adsorption and endothelial cell growth on the studied surfaces (Fig. 27). As a consequence, the fibronectin adsorption seems to be suitable to make predictions about cell proliferation.

3.3
Immobilization of Cell Adhesion Mediators on Biocompatible Polymer Surfaces to Improve Cell Seeding

A specific demand for a biomaterial surface is its compatibility with organ specific cells, which leads to the approach to cell seeding before implantation. Therefore, implants for blood contact may be durably covered with a monolayer of endothelial cells. Thus, a suitable interface between the synthetic carrier material and the individual cells is required [132]. For this reason, a concept based on covalent coupling of cell adhesion promoters, like fibronectin or the pentapeptide GRGDS belonging to the endothelial cell binding domain of fibronectin, at the biomaterial surface is pursued. To achieve this the base polymer has

Table 2. Selected carrier polymers for immobilization of cell adhesion mediators

		(PCU)	(Tecoflex™)
Polyurethane	Diol	 Polyhexamethylene-carbonate diol	 Poly(oxytetra-methylene)
	Diisocyanate	 4,4'-Diphenylmethane-diisocyanate	 4,4'-Dicyclohexyl-methanediisocyanate
	Chain extender	1,4-Butanediol	1,4-Butanediol
Poly(vinyl chloride)	Poly[(ethene-co-vinyl acetate)-co-carbon mon-oxide-graft-vi-nyl chloride] (PVC/EVACO)		
Polydimethyl-siloxane	Sylgard 184 two-component elastomer		

to be functionalized. Often it is desirable to first attach a suitable spacer molecule before the cell mediator is coupled (Fig. 28).

Poly(ether urethanes) are of special interest due to their good mechanical properties, broad synthetic possibilities, and rather good blood compatibility for medical applications, e.g. catheters, heart valves [47]. A poly(carbonate urethane) (PCU) comprising poly(hexamethylene carbonate), the aromatic diisocyanate 4,4'-diphenylmethane diisocyanate (MDI) and 1,4-butanediol as chain prolonger has been developed (Table 2) [133].

To introduce functional groups, polymer blends were produced in the extruder from PCU and poly(vinyl alcohol) (PVA) with 10, 20 and 30 mass% PVA. In addition, the commercial products Mitrathane™ and Tecoflex™ were applied as carrier materials. While Mitrathane is built up of the aromatic diisocyanate MDI, hydroxytelechelic polytetrahydrofuran and an aliphatic amine as chain prolonger, Tecoflex™ consists of a cycloaliphatic diisocyanate component (Table 2). The

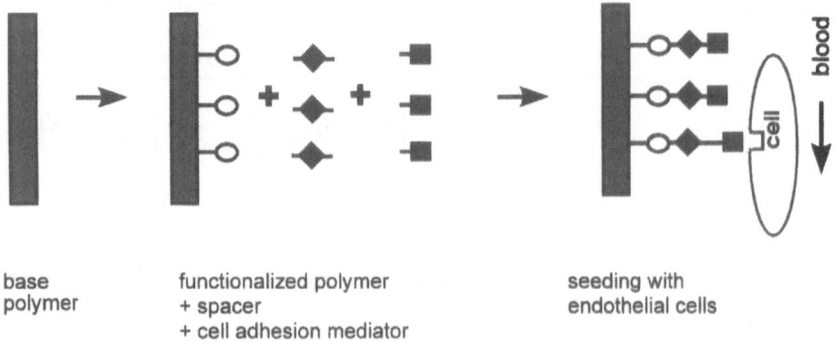

base   functionalized polymer   seeding with
polymer   + spacer   endothelial cells
                    + cell adhesion mediator

Fig. 28. Schematic illustration of an interface for endothelial cell seeding

functionalization of Mitrathane was carried out by vinyl acetate plasma treatment and saponification of the grafted vinyl acetate groups while the functionalization of Tecoflex was carried out by means of plasma-induced graftpolymerization of 4-(acryloyloxy)butylhydrogenglutarate [134].

Another carrier polymer is polymer-plasticized poly(vinyl chloride) (PVC/EVACO), which was developed for medical application. Hydroxyl groups as functional groups were created through saponification of the vinyl acetate groups of the EVACO component.

In ophthalmology poly(dimethyl siloxane) requires good cell seeding, which is not possible on extremely hydrophobic polymer surfaces. Due to the need for reactive groups on the carrier polymer surface, functional groups are built up by plasma-induced graftpolymerization of suitable monomers like HEMA.

The spacer molecules guarantee the accessibility of the immobilized substances [135]. In the selection of suitable spacer molecules attention has to be paid to the following points:

1. Cyclization reactions must be avoided.
2. The length of the spacer has to be sufficiently large.
3. The immobilization of the protein or peptide sequence has to be achieved in such a way that the bioactive group is exposed to the surface.

Some spacer molecules and the corresponding carrier polymers are shown in Table 3. Benzoquinone was bound to the surface of a poly(carbonate urethane) modified with poly(4-hydroxybutyl acrylate) (PCU/HBA) and to the vinyl acetate grafted and saponified poly(ether urethane) (Mitrathane) (PEU/VA-plasma) [136–139].

Aliphatic dicarboxylic acid dichlorides were applied to the surface-saponified blend of 50 wt% poly(ethylene-*co*-propene) and 50 wt% poly(ethylene-*co*-vinyl acetate) (PP/EVA). After coupling without cyclization, the second carboxylic acid group is available for the covalent binding of the protein via the *N*-terminal amino group [140]. The use of anhydrides, e.g. glutaric acid anhydride, is another possibility to introduce free carboxyl groups into the polymer surface.

Table 3. Bifunctional spacers for activation of polymer surfaces

Spacer molecule	Name	Carrier polymer
	Benzoquinone	PEU/VA plasma
	Succinyl dichloride	PCU/PHBA
	Oxalyl dichloride (gas phase)	PP/EVA
	4-Isocyanatomethyl butanoate	PCU/PVA
	Glutaric acid anhydride	PHBA
	Hexamethylene diisocyanate	PVC/EVA

The use of hexamethylene diisocyanate and 4-isocyanatomethyl butanoate has the advantage of high reactivity of the isocyanate function and high hydrolytic stability of the urethane group compared with the ester bond.

3.3.1
Immobilization of GRGDS on Modified Polyurethanes

The necessary functional groups to activate the surface with the spacer molecule benzoquinone on the Mitrathane surface [141] were introduced using plasma polymerization of vinyl acetate followed by saponification. Besides the immobilization of single amino acids, the fragment of fibronectin GRGDS as cell adhesion mediator was immobilized [142,143].

Equation (6) shows the modification steps of the polymer blend consisting of poly(carbonate urthane) and poly(vinyl alcohol) (PCU/PVA). No direct proof of OH functions at the PCU/PVA blend surface was achieved; however, after introduction of the trifluoroacetyl reporter group, XPS showed the presence of 2.5 atom% fluorine. The OH functions of the PCU/PVA blend surface were first reacted with the bifunctional spacer 4-isocyanatomethyl butanoate. Due to the different reactivity of the functional groups of the spacer, the coupling with the polymer surface occurs exclusively through the isocyanate groups. After saponification of methyl ester functions the carboxylic groups are available for further reactions (Eq. (6)).

$$(6)$$

XPS was used to prove the coupling of 4-isocyanatomethyl butanoate with the PCU/PVA blend surface and the saponification step. The high resolution C_{1s} and O_{1s} spectra of the unmodified surface and of the surface after reaction with phenylisocyanate are shown in Fig. 29 for comparison. The photoline at 289.3 eV is assigned to the carbon in the urethane group as well as in the ester group. The photoline at 289.8 eV corresponds to the carbon in the urethane group obtained by reaction with phenylisocyanate. No resolution of the photoline for the ester and urethane groups is observed because of the shift of the C_{1s} photoline of the ester group to the lower binding energies [144].

After saponification of the ester function the carboxyl function exists as sodium carboxylate. The C_{1s} photoline of the carbon of the sodium carboxylate group shows up at a binding energy of 288.4 eV [145]. In addition, a photoline is present in the O_{1s} elemental spectrum at 535.8 eV, which may be due to adsorbed water [146].

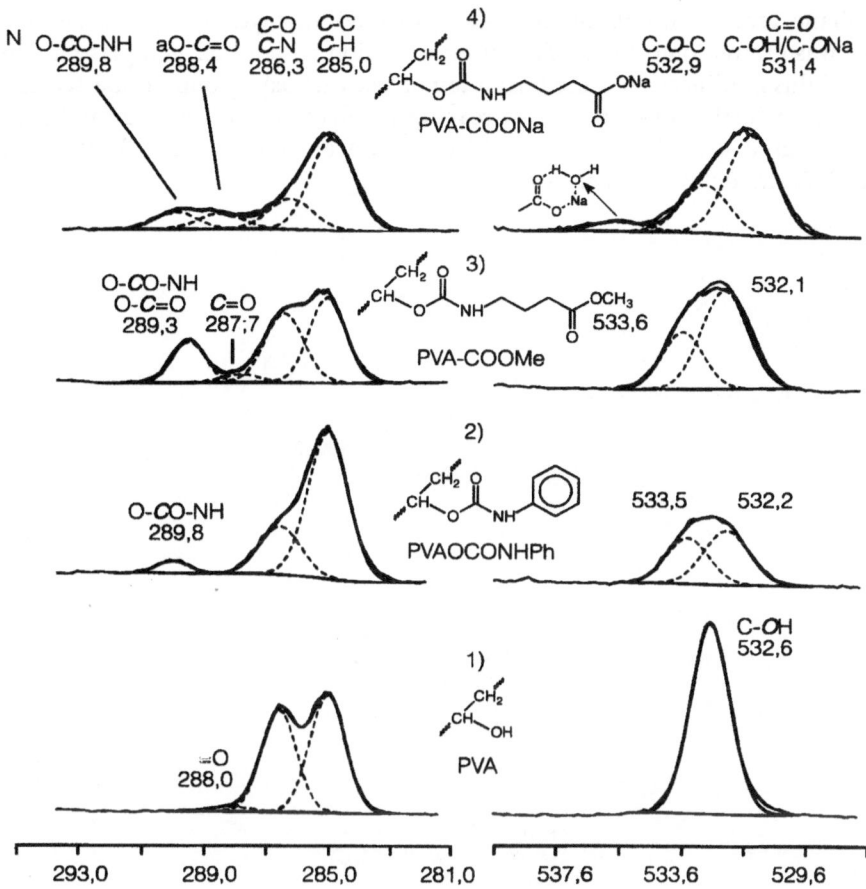

Fig. 29. Spectra from C_{1s} (*left*) and O_{1s} (*right*) XPS of unmodified PVA (*1*), surface modified with phenylisocyanate (*2*), surface modified by spacer coupling (*3*) and after hydrolyzation (*4*)

Since the result of the saponification of the methyl ester of immobilized 4-iso-cyanatomethyl butanoate was not proven directly by XPS the free acid group was reacted with 4-amino-TEMPO using the carbodiimide method. The immobilized 4-amino-TEMPO leads to an ESR signal characterized by reduced molecular motion in comparison with an uncoupled spin probe as a control sample [133]. Consequently, the modified PCU/PVA surface is suitable for the immobilization of the fibronectin fragment GRGDS via the N-terminus [147]. A pepide content of 8 to 15 nmol · cm^{-2} on the PVC/PVA surface immobilized with GRGDS was verified by means of amino acid analysis.

Tecoflex as Carrier Material. Tecoflex was chosen as an appropriate carrier material due to the fact that it does not contain any aromatic components and it is

used in medical applications for blood contact. In order to modify the polyurethane Tecoflex, the monomer 4-(acryloyloxy)butylhydrogen glutarate (AB-
HG) was grafted onto the surface after activation with an oxygen plasma. The
aim of this modification is the formation of free carboxyl groups at the Tecoflex
surface for further immobilization reactions. Poly(hydroxybutyl acrylate) (PH-
BA) was used as model surface [146]. The modification steps of the Tecoflex and
the PHBA model surfaces are summarized in Eq. (7).

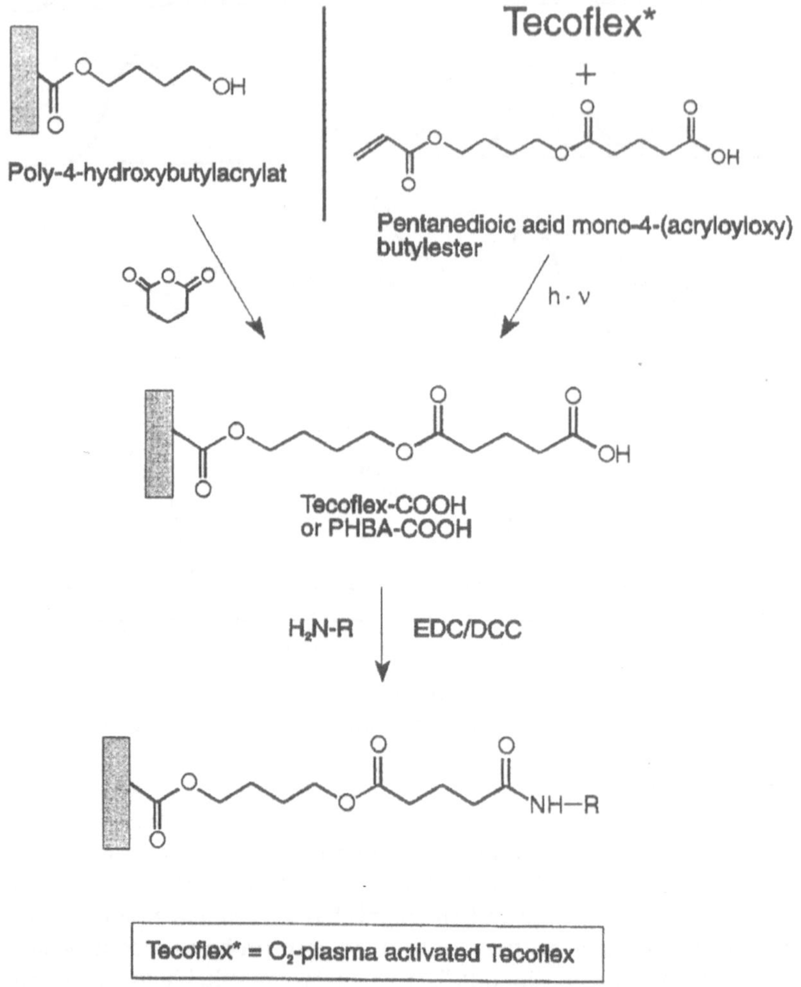

Poly-4-hydroxybutylacrylat

Tecoflex*

Pentanedioic acid mono-4-(acryloyloxy)
butylester

h · v

Tecoflex-COOH
or PHBA-COOH

(7)

H₂N-R   EDC/DCC

Tecoflex* = O₂-plasma activated Tecoflex

The polyABHG chains grafted onto the Tecoflex surface assume the function
of a spacer and present free carboxylic groups. In order to achieve similar structures at the model surface, PHBA was reacted with glutaric acid anhydride as a
spacer molecule. The covalent bond between the glutaric acid anhydride and the
PHBA surface was proven using SIMS [146]. The free carboxylic groups on the

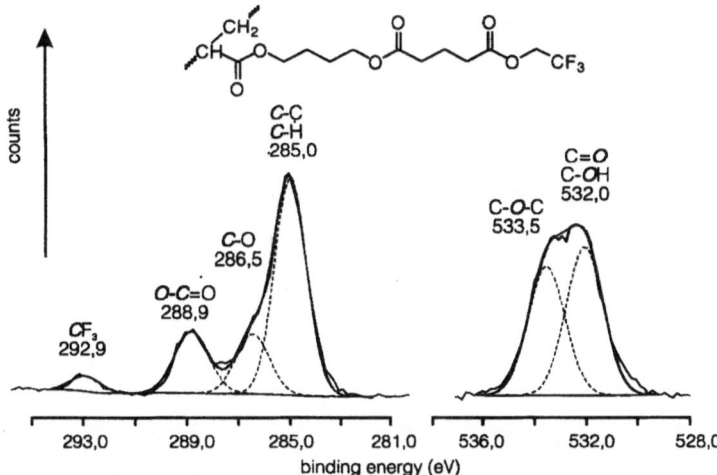

Fig. 30. Spectra from C_{1s} (*left*) and O_{1s} (*right*) XPS of the PHBA/COOH surface after reaction with trifluoroethanol

PHBA surface were detected by means of XPS after reaction with trifluoroethanol using the carbodiimide method. The high resolution C_{1s} and O_{1s} spectra are presented in Fig. 30. The C_{1s} photoline of the CF_3 group at 292.9 eV proves the coupling of the trifluoroethyl reporter group. A fluorine content of 2.9 atom% compared with the expected maximal value of 4.3 atom% (yield=100%) suggests a yield of 67%.

ESR spectroscopy is a suitable method to follow the immobilization of 4-amino-TEMPO on free carboxylic groups (Fig. 31) [133]. The ESR spectrum of the immobilized spin label is clearly distinguished from the spectrum of the adsorbed spin label (reference) [133]. In full analogy to the coupling of the spin label, the coupling of GRGDS was performed in aqueous medium using EDC as coupling reagent. The immobilization was proved by the detection of characteristic fragments by means of SIMS.

The immobilization of the peptide GRGDS on the PHBA model surface – after reaction with glutaric acid anhydride – was indicated by IR and XPS [146]. MS showed a fragment of m/c=430, which corresponds to Arg-Gly-Asp-Ser. The benzylic group introduced as protecting group in the C-terminus of serine and in the side chain of asparagine results in the formation of the tropylium cation which shows up at m/c=91 only after immobilization of the protected GRGDS, and not after a blank experiment involving adsorption (Fig. 32).

Comparative proliferation studies with human endothelial cells were carried out to investigate the biological activity of GRGDS immobilized at the Tecoflex™ surface. Contrary to expectations, after seven days of proliferation a decrease in cell growth was observed compared with the value obtained on Tecoflex surface reacted with the spacer molecule [146]. A possible explanation is that the interaction of cells with the peptide sequence GRGDS by means of integrines is hin-

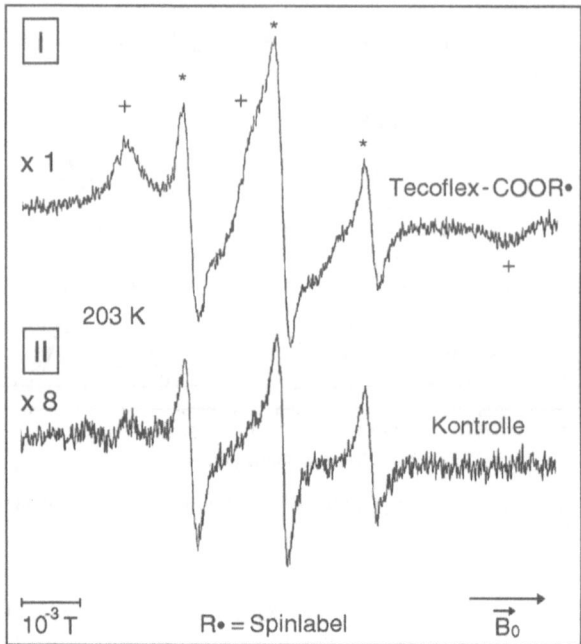

Fig. 31. ESR signals of Tecoflex modified with pentanedioic acid mono-4-(acryloy-loxy)butyl ester and 4-amino-TEMPO in comparison with a control with physiosorbed un-coupled spin probe as control sample

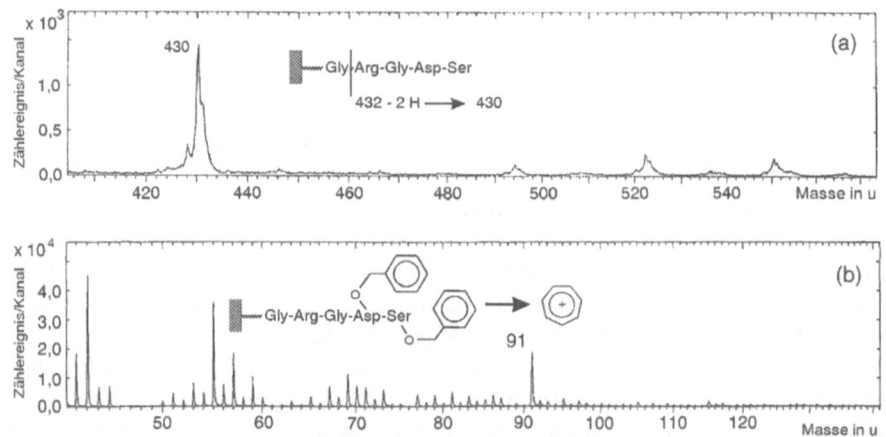

Fig. 32. Positive secondary ion mass spectrum of immobilized GRGDS (*a*) and of the immo-bilized benzylic ester of the GRGDS (*b*) on a PHBA/COOH surface

dered by a protein film which is adsorbed upon contact with the biomedium. It is known that protein adsorption is faster than cell adhesion and proliferation [148]. For this reason, the covalent binding of the whole protein fibronectin on selected carrier polymers is sought. Free accessibility of the endothelial cell binding domain of the immobilized fibronectin has to be guaranteed to achieve good cell adhesion and proliferation on such a surface.

3.3.2
Immobilization of Fibronectin on Medical Grade Poly(vinyl chloride)

The grafted polymer poly[(ethylene-*co*-vinyl acetate-*co*-carbon monoxide)-*graft*-vinyl chloride] (PVC/EVACO) known in medical applications as plasticized poly(vinyl chloride) was applied as a carrier polymer for the covalent immobilization of fibronectin. The surface modifications were carried out on foils both with closed surface structure and with microporous surfaces produced by a phase-inversion technique. The vinyl acetate groups of the carrier polymer were saponified and then reacted with hexamethylene diisocyanate (HDI) as a spacer. Fibronectin is immobilized upon reaction of the free isocyanate group with an amino group of the protein (Fig. 33). The saponification of the carrier polymer was verified by IR-ATR and XPS [149]. The presence of hydroxylic groups after hydrophilization is demonstrated by a contact angle of 61° while that of the basic polymer is 110°.

The coupling of HDI (5 d in diethyl ether at 250 °C) was proven by means of XPS, indicating a nitrogen content of 3.7 atom% and the urethane carbon. IR-ATR spectra show the characteristic absorption bands of the free isocyanate

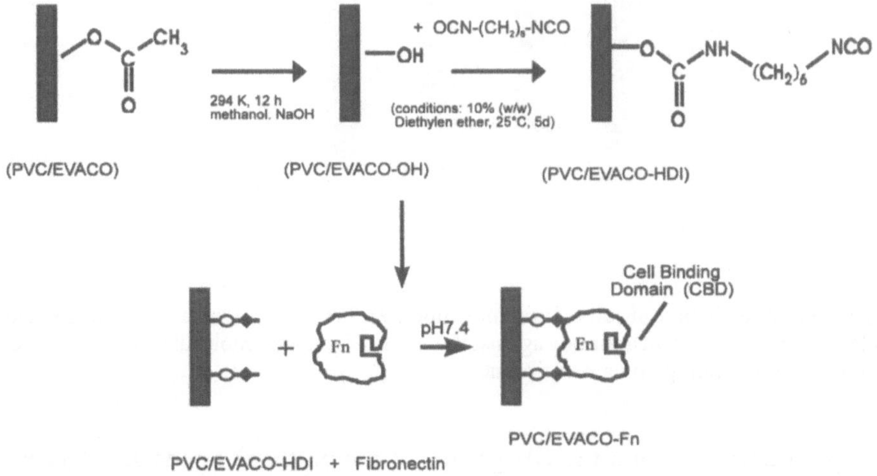

Fig. 33. Schematic illustration of the modification steps of immobilization of fibronectin onto the carrier polymer PVC/EVACO

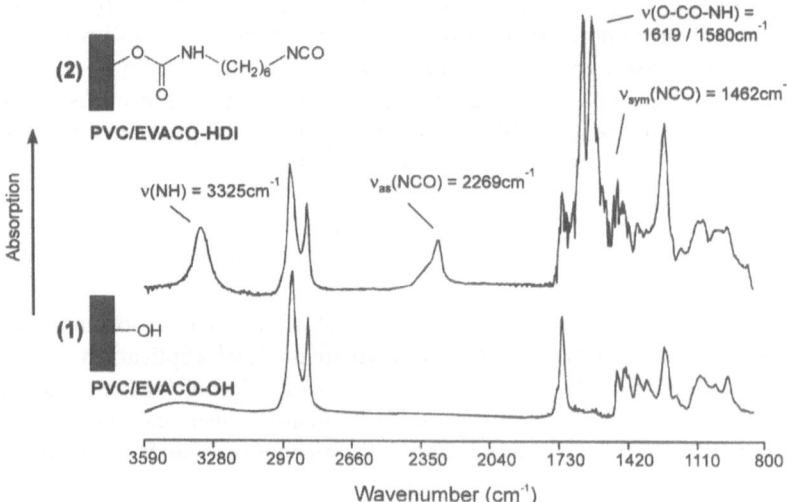

Fig.34. IR-ATR spectra of the saponified PVC/EVACO surface (*1*) and the saponified surface after coupling with the spacer hexamethylene diisocyanate (HDI) (*2*)

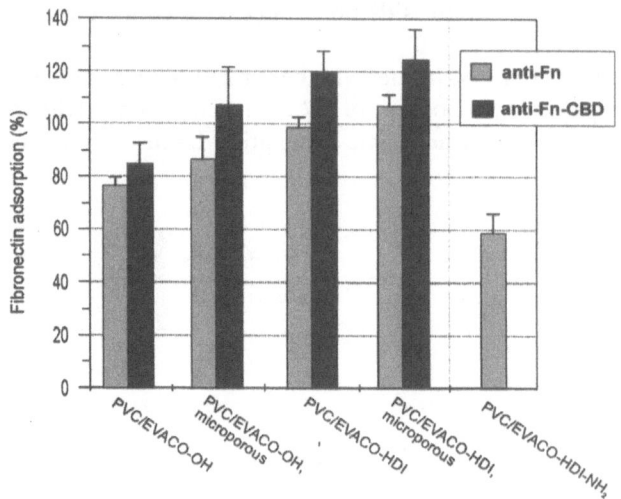

Fig.35. Fibronectin binding onto differently modified PVC/EVACO surfaces by means of an ELISA technique using an antibody against the whole fibronectin molecule and an antibody against the cell binding domain of fibronectin

groups at 2269 cm^{-1} and the NH groups of the urethane bonds at 3328 cm^{-1} (Fig. 34). The use of photometric spectroscopy allows the evaluation of the accessibility of the free isocyanate groups by fluorescence labeling with *p*-fluoroaniline. The qualitative judgment of the penetration depth of the fluorescent la-

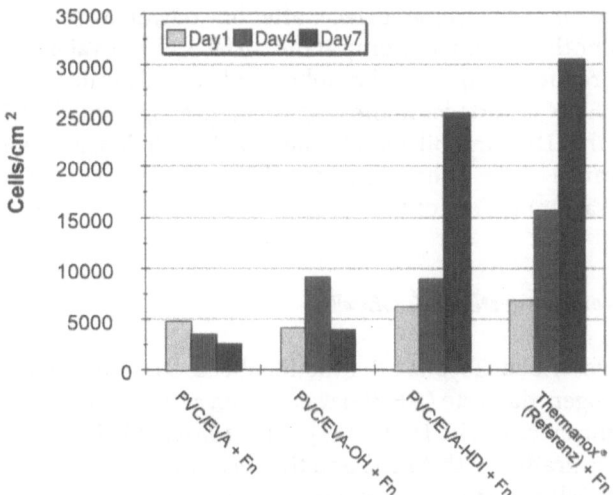

Fig. 36. Proliferation behavior of human endothelial cells on modified PVC/EVACO surfaces and the reference surface Thermanox (n=3, cell number determined by a Coulter counter)

bel [150] points to a thickness of the modified layer with free isocyanate groups in the µm range.

The covalent binding of fibronectin (10 µg/ml) was carried out in aqueous medium at pH 7.4. The amount of immobilized fibronectin was measured semiquantitatively by means of an ELISA technique. The values of fibronectin binding (via covalent bonds or adsorption) presented in Fig. 35 are given with reference to the TCPS surface.

The amount of fibronectin at microporous surfaces is as expected generally higher than at smooth ones. The amount of fibronectin at the surfaces activated with HDI overtakes the values of the non-activated surfaces, e.g. the amount of fibronectin of 106% at an activated microporous surface is up to 20% higher than at a non-activated, saponified microporous surface. This increase is attributed to the covalently bound amount of fibronectin in addition to the adsorbed portion. The relatively high fibronectin binding capacity of 75% at the saponified, smooth PVC/EVACO surface is accounted for by its favorable contact angle (61°). The "de-activation" of the free isocyanate functions at the PVC/EVACO surface by means of carbonate buffer at pH 9.6 (PVC/EVACO-HDI-NH$_2$) leads to a strong decrease in the amount of adsorbed fibronectin. Studies with structure-specific monoclonal antibodies point to the fact that the accessibility of the cell binding domain of the fibronectin molecule at the PVC/EVACO surfaces is fully retained.

The effect of the modified PVC/EVACO surface on the proliferation behavior of human endothelial cells was investigated by means of endothelial cell seeding studies after 1, 4 and 7 d (Fig. 36). While the endothelial cells show similar proliferation behavior at all sample surfaces after one day, an increase in cell density is observed after the seventh day only at the surface with covalently immobilized fibronectin. A reduction in the cell number is observed at the untreated and saponified PVC/EVACO surfaces in the long-term investigation (7 d). In the case of the surface with HDI-immobilized fibronectin, the cell density after the seventh day is in the range of the reference material Thermanox, which proves the efficacy of the covalent binding of fibronectin.

3.3.3
Immobilization of Fibronectin on Tecoflex™

The already described Tecoflex™ functionalized by grafting of 4-(acryloyloxy)butylhydrogen glutarate (ABHG) was chosen as a second carrier polymer to immobilize fibronectin via HDI [151]. The untreated Tecoflex surface, the Tecoflex surface grafted with ABHG and the Tecoflex surface activated with HDI were comparatively investigated using IR-ATR spectroscopy with an information depth of approximately 1 µm (Fig. 37). The grafting with ABHG on the polymer surface is proved by the absorption bands at 1162 and 1731 cm^{-1} due to the v(C-O-C) and the v(C=O) band of ABHG. The HDI conversion of the functionalized Tecoflex surface is demonstrated by the characteristic IR absorption band due to the terminal isocyanate groups at 2271 cm^{-1} and the NH groups of the urethane bonds at 3325 cm^{-1}. In addition, the coupling of the spacer with the polymer surface is confirmed by means of XPS with results for the elemental composition of the surface activated with HDI of a nitrogen content of 10 atom%.

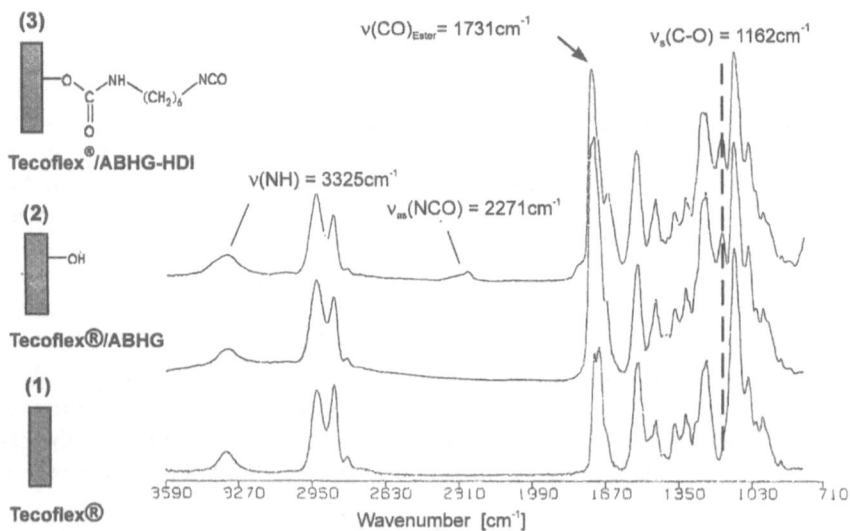

Fig. 37. IR-ATR spectra of the unmodified (*1*), ABG*H*-grafted (*2*) and HDI-activated (*3*) Tecoflex surface

The immobilization of fibronectin at pH 7.4 was semiquantitatively determined by means of an ELISA technique. In Fig. 38 the amounts of fibronectin, with respect to the reference surface TCPS adsorbed at an untreated Tecoflex surface, a Tecoflex surface grafted with ABHG and the Tecoflex surface activated with HDI are presented.

The proliferation of human endothelial cells was indirectly determined by measurement of the protein content after 1, 4 and 7 d using the sulforhodamine B method (SRB method) [152]. The proliferation rates were determined in a serum-free culture medium, i.e. without addition of fibronectin (Fig 39). While any significant difference in cell growth behavior was not observed at the modified Tecoflex surfaces after one day of proliferation, a remarkable increase in the proliferation of human endothelial cells was seen after seven days only at the surface with immobilized fibronectin. These results confirm the biological activity of fibronectin bound to Tecoflex, especially the free accessibility of the cell binding domain as a main requirement for good cell seeding. It is not generally observed that upon unspecific immobilization of a bioactive polymer its biological activity is retained. In the case of hirudin, for example, a differentiating protecting group procedure has to be chosen to obtain biological activity after immobilization.

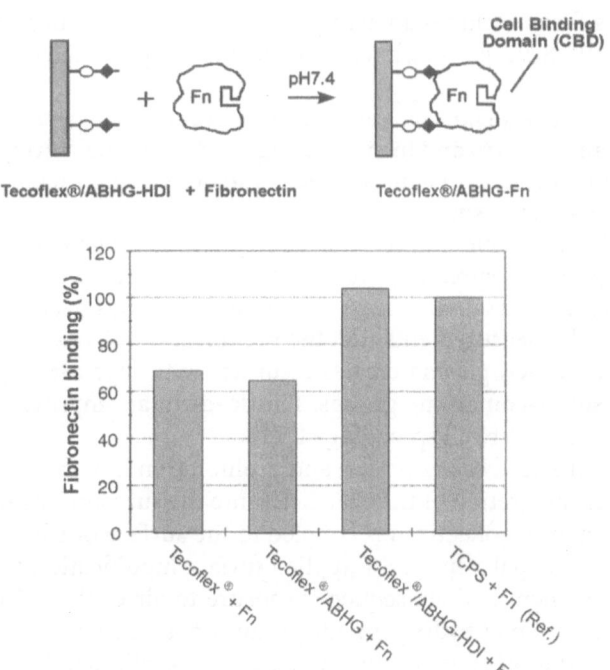

Fig. 38. Fibronectin binding onto differently modified Tecoflex surfaces by means of an ELISA technique using an antibody against the whole fibronectin molecule

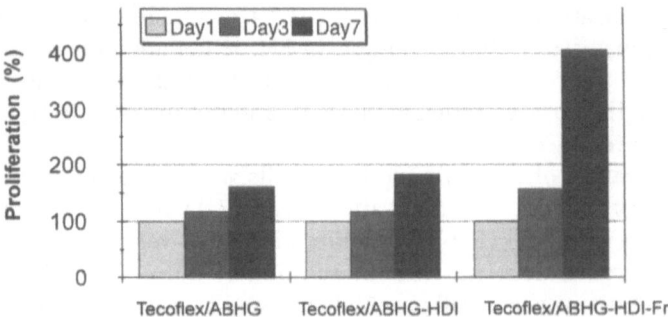

Fig. 39. Proliferation behavior of human endothelial cells on modified Tecoflex surfaces in serum-free culture medium (protein content by means of the SRB method)

3.3.4
Surface-Modified Silicone for Ophthalmological Applications

Silicone rubber has been in clinical use since 1959 because of its inertness, flexibility, optical qualities and gas permeability. It is mainly used for the replacement of soft tissues. In ophthalmology, medical grade silicone has been used routinely in retinal detachment surgery and for the past 12 years as intraocular lenses. Today's focus includes an artificial cornea with an optical part made of silicone. Recently, Schrage et al. have developed a one-piece silicone keratoprosthesis [153].

The medical grade silicone rubber Sylgard 184 consists of a crosslinked polydimethylsiloxane. In vitro and in vivo testing has proven the non-cytotoxicity in direct and indirect contact. There is no ocular toxicity and no acute foreign body reaction in contact with soft tissues.

Due to its hydrophobic nature, silicone is prone to a fairly high protein adsorption and poor cell spreading. To alter these undesired properties, it is necessary to generate functional groups on the surface. It has been shown by XPS and contact angle measurements that the modification of silicone by means of treatment with an SO_2 plasma creates a surface with increased hydrophilicity and oxidized sulfur-containing groups. This leads to an improved adhesion of collagen IV and a better cell spreading [154].

To achieve enhanced cell adhesion and proliferation in vitro which probably leads to a better integration of the prosthesis into the surrounding tissue the cell adhesion mediator fibronectin was coupled to the surface of silicone. Figure 40 shows the two principal steps of the applied surface modification procedure. Argon plasma treatment and subsequent exposure to air of the silicone samples leads to the formation of hydroperoxide groups on its surface which are used to initiate polymerization of acrylic acid (AAc), methacrylic acid (MAAc) and glycidyl methacrylate (GMA) generating carboxylic and epoxy groups, respectively.

High-resolution XPS of carbon proves the grafting. The percentage of silicium decreases from 22 to 10 atom%. The carbon content increases from 47 to 64

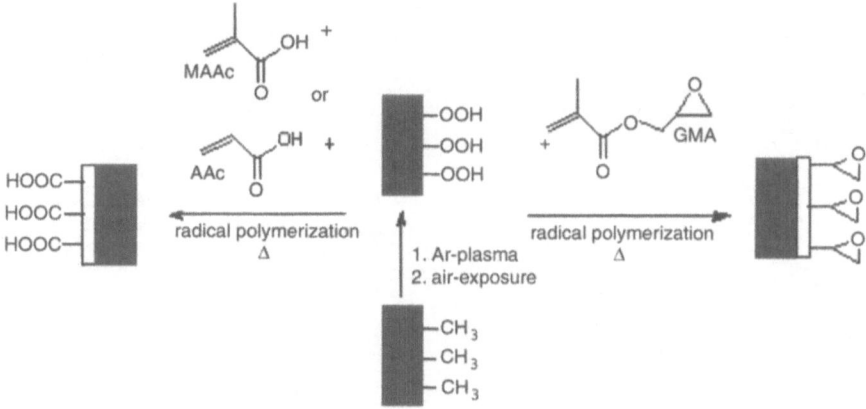

Fig. 40. Schematic illustration of plasma-induced graftcopolymerization of hydrogels and GMA on silicone surface

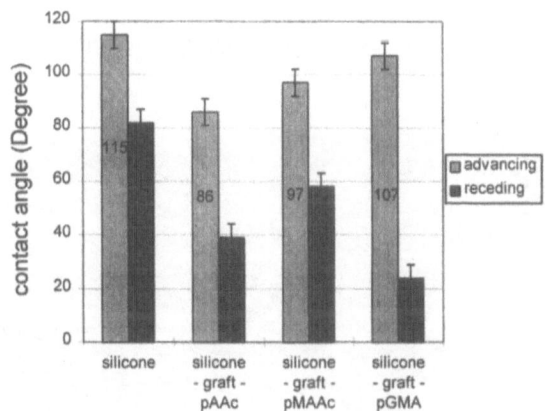

Fig. 41. Contact angles (Wilhelmy plate method) of untreated and differently modified silicone surfaces

atom%. The surface composition of the grafted sample coincides with the surface composition of the copolymer.

The advancing and receding contact angles for untreated silicone, silicone-*graft*-PAA, silicone-*graft*-PAAc and silicone-*graft*-PGMA are shown in Fig. 41. Fibronectin is reacted with the epoxy groups introduced by Ar plasma induced grafting. The amount of fibronectin immobilized on the surface was measured by means of a sandwich-ELISA using a structure-specific primary antibody against the cell binding domain of fibronectin. The results are given in Fig. 42 (untreated silicone is set at 100%).

Silicone grafted with acrylic acid or methacrylic acid exhibits a hydrogel character at its surface showing less protein adsorbed than on untreated sili-

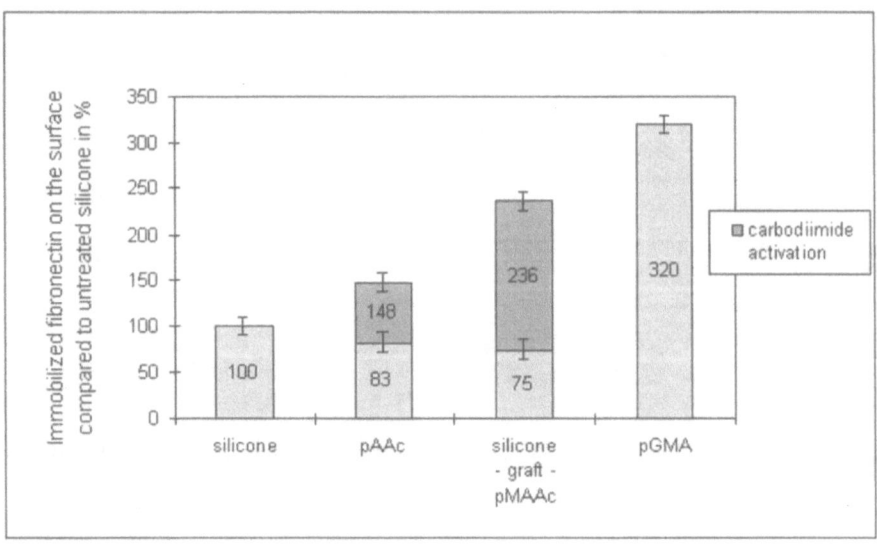

Fig. 42. Fibronectin binding of untreated and differently modified silicone surfaces

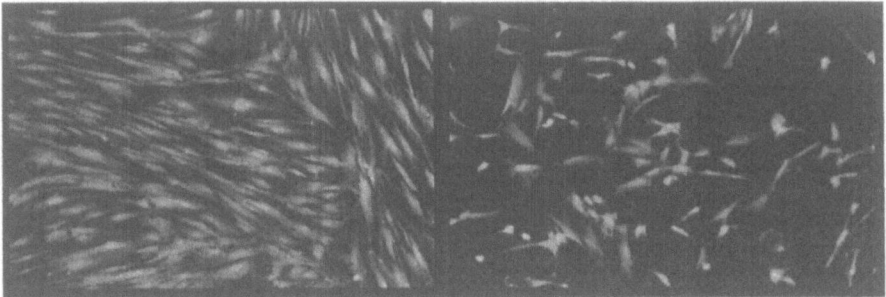

Fig. 43. Proliferation of human corneal fibroblasts on silicone surface with immobilized fibronectin (*left*) in comparison with an unmodified surface (*right*)

cone. Activation of the carboxylic acid groups by carbodiimides leads to an increase of surface-bound fibronectin. This effect is even enhanced for silicone-*graft*-PGMA.

In vitro testing shows that the vitality and morphology of L929 cells and human corneal fibroblasts is highly increased on the surface of modified silicone with immobilized fibronectin compared to untreated silicone and silicone with adsorbed fibronectin and compares well with the control (Thermanox) (Fig. 43). The results show that silicone is a well-suited polymer for biomedical applications and especially for ophthalmology. The problem of its hydrophobicity and lack of functional groups can be solved by applying the above-mentioned surface modification strategy. Covalent immobilization of cell adhesion mediators leads to an improved cell response.

4
Discussion

A large variety of polymers are used for medical applications in direct contact with the biosystem, short and long term, intra- and extracorporal. The polymers generally fulfill the mechanical requirements perfectly; the interfacial biocompatibility, however, is imperfect. This is mainly due to the fact that it is not known how a foreign body surface has to be designed to exert perfect biocompatibility. The present article shows that by well-defined surface modification procedures the response of single parameters of the biosystem may be controlled. This is certainly not the solution to obtain a biocompatible material. It is, however, a first step to increase our knowledge of the interaction between an artificial material and the most complex biosystem.

References

1. Ratner BD (1989) Biomedical applications of synthetic polymers. In: The synthesis characterization reaction and application of polymers. Pergamon Press, chap 7, p 201
2. Ratner BD (1993) J Biomed Mat Res 27:837
3. Williams DF (1987) Definitions in biomaterials. In: Progress in biomedical engineering. Elsevier, Amsterdam
4. Williams DF (1990) Concise encyclopedia of medical and dental materials. Pergamon Press
5. Szycher M (1991) High performance biomaterials. In: A comprehensive guide to medical and pharmaceutical applications. Technomic, Lancaster
6. Blass CR (1992) Med Dev Tec 33:32
7. Williams DF (1994) J Mat Sci 5:303
8. National Toxicology Program Technical report on the Carcinogenesis Bioassay of 4,4'-Methylenedianiline Dihydrochloride, NTP-81-143, NIH Publication No. 82-2504, National Institutes of Health, Betheseda, Maryland (1982)
9. Klinkmann H, Falkenhagen D, Courtney JM, Gurland HJ (eds) (1987) Uremia therapy. Springer, Berlin Heidelberg New York
10. Williams DF (1990) Interdisk Sci Rev 15:20
11. Ikada Y, Shalaby W. et al. (eds) (1994) ACS Symp Series No. 540
12. Wegner G (1985) Mehrphasige Polymere. In: Batzer H (ed) Polymere Werkstoffe. Thieme Verlag, Stuttgart, p 679
13. Wirsching J (1997) PhD thesis, Technical University Aachen, Shaker Verlag Aachen
14. Castner DG, Ratner BD, Hoffman AS (1990) J Biomater Sci Polymer Edn 1:191
15. Grainger DW, Okano T, Kim SW, Castner DG, Ratner BD Briggs D, Sung YK (1990) J Biomed Mat Res 24:547
16. Tyler BJ, Ratner BD, Castner DG, Briggs D (1992) J Biomed Mat Res 26:273
17. Ratner BD (1987) Surface contamination and biomaterials. In: Mittal KL (ed) Treatise on clean surface technology. Plenum Press, New York, p 247
18. Ziats NP, Miller KM, Anderson JM (1988) Biomaterials 9:5
19. Bakker D, van Blitterswijk CA, Hesseling SC, Grote JJ, Daems WT (1988) Biomaterials 9:14

20. Williams DF (1987) J Mater Sci 22:3421
21. Dankert J, Hogt AH, Feijen J (1986) In: Williams DF (ed) CRC critical reviews in biocompatibility, p 219
22. Courtney JM, Lamba NMK, Sundaram S, Forbes CD (1994) Biomaterials 15:737
23. Salzman EW (1986) Blood material interaction In: Interaction of the blood with natural and artificial surfaces, Dekker Inc, New York, p 39
24. Meyer JG (1986) Blutgerinnung und Fibrinolyse, Deutsche Ärzte Verlag, Köln
25. Baszkin A (1986) The effect of polymer surface composition and structure on adsorption of plasma proteins. In: Dawids S, Bantjes A (eds) Blood compatible materials and their testing. Martinus Nijhoff Publishers, Dortrecht, p 39
26. Brash JL (1983) Mechanism of adsorption of proteins to solid surfaces and its relationship to blood compatibility. In: Szycher M (ed) Biocompatible polymers, metals and composites. Technomic, Lancaster, p 35
27. Kottke-Marchant K, Anderson JM, Umemura Y, Marchant RE (1989) Biomaterials 10:147
28. Ikada Y (1984) Adv Polym Sci 57:103
29. Andrade JD (1973) Med Instrum 7:110
30. Baier RE, Gott VL, Feruse A (1970) Trans Amer Soc Artif Intern Organs 16:50
31. Ratner BD, Hoffman AS, Hanson SR, Harker LA, Whiffen JD (1979) J Polym Sci Polym Symp 66:363
32. Bantjes A (1978) Brit Polym J 10:267
33. Bruck SD (1979) J Polym Sci Polym Symp 66:283
34. Sawyer PN, Srinivasan S (1967) Am J Surg 114:42
35. Norde W, Lyklema J (1985) Proton titration and electrokinetic studies of adsorbed protein layers. In: Andrade JD (ed) Surface and interfacial aspects of biomedical polymers. Plenum Press, New York, p 241
36. Szycher M (1983) Thrombosis hemostasis and thrombolysis at prosthetic interfaces. In: Szycher M (ed) Biocompatible polymers, metals and composites. Technomic, Lancaster
37. Yu J, Sundaram S, Weng D, Courtney JM, Moran CR, Graham NB (1991) Biomaterials 12:119
38. Beissinger RL, Leonard EF (1981) Trans Amer Soc Artif Intern Organs 27:225
39. Sonderquist ME, Walton AG (1980) J Colloid Interface Sci 75:386
40. Lundström I, Elwing H (1990) J Colloid Interface Sci 136:68
41. Vroman L (1987) Semin Thromb Hemostasis 13:79
42. Vroman L (1987) Ann NY Acad Sci 516:300
43. Garbassi F, Morra M (1994) Biomedical materials. In: Ochiello E (ed) Polymer surfaces, from physics to technology. Wiley, Chichester, p 395
44. Hess F, Jerusalem C, Braun B, Grande P (1984) New formation of the neointima in the polymeric artifical vessels. In: Planck H, Egbers G, Syre I (eds) Polyurethane in biomedical engineering. Elsevier, Amsterdam, p 301
45. Alberts B, Bray D, Lewis J, Raff M, Roberts K, Watson JD (1990) Molekularbiologie der Zelle. VCH, Weinheim
46. Wachem PB (1987) Interaction of cultured human endothelial cells with polymeric surfaces, PhD thesis, University of Twente, Netherlands
47. Klee D, Höcker H, Mittermayer C (1987) Plastverarbeiter 38/4:54; 38/5:76; 38/6:52
48. Pierschbacher MD, Hayman EG, Ruoslahti E (1983) Proc Natl Acad Sci USA 80:1224
49. Pierschbacher MD, Ruoslahti E (1984) Nature 309:30
50. Pierschbacher MD, Ruoslahti E (1984) Proc Natl Acad Sci USA 81:5985
51. Ruoslathi E, Pierschbacher MD (1986) Cell 44:517
52. Müller G, Gurrath M, Kessler H, Timpl R (1992) Angew Chem 104:341
53. Lin HB, García-Echeverría C, Asakura S, Sun W, Mosher DF, Cooper SL (1992) Biomaterials 13:905
54. Sentissi JM, Ramberg K, O'Donnell RJ, Connolly RJ, Callow AD (1986) Surgery 99:337

55. Lin HB, Sun W, Mosher DF, García-Echeverría C, Schaufelberger K, Lelkes PI, Cooper SL (1994) 28:329
56. Ratner BD (1985) Graft copolymer and block copolymer surfaces. In: Andrade JD (ed) Surface and interfacial aspects of biomedical polymers, Vol: Surface chemistry and physics. Plenum Press, New York, p 373
57. Yui N, Kataoka K, Sakurai Y, Akutsu T (eds) (1986) Microdomain-structured polymers as antithrombogenic materials in artificial heart. Springer, Berlin Heidelberg New York
58. Okano T, Nishiyama S, Shinohara I, Akaike T, Sakurai Y, Kataoka K, Tsuruta T (1981) J Biomed Mat Res 15:393
59. Lelah MD, Cooper SL (1986) Polyurethanes in medicine. CRC Press, Boca Raton, Florida
60. Sharma CP, Szycher M (1991) Blood compatible materials and devices. Technomic, Lancaster
61. Leake DL, Cenni E, Cavedagna D, Stea S, Ciapetti G, Pizzoferrato A (1989) Biomaterials 10:441
62. Matsuda T (1989) Nephrol Dial Transplant 4:60
63. Lee JH, Lee HB, Andrade JD (1995) Prog Polym Sci 20:1043
64. Andrade JD (1976) Hydrogels for medical and related applications. ACS Symp Series
65. Hoffmann AS (1984) Adv Polym Sci 57:101
66. Forbes CD, Courtney JM (1994) Thrombosis and artificial surfaces. In: Bloom AL, Forbes CD, Thomas DP, Tuddenham EGD (eds) Haemostasis and thrombosis. Churchill Livingstone, Edinburgh, p 1301
67. Engbers GH, Feijen J (1992) Int J Artif Org 14:199
68. Chapman D, Charles SA (1992) Chem Br 28:253
69. Hayward JA, Durrani AA, Shelton CJ, Lee DC, Chapman D (1986) Biomaterials 7:126
70. Jerg KR, Emonds M, Müller U, Keller R, Baumann H (1991) ACS Polym Preprints 32:243
71. Ratner BD (1988) The surface characterization of biomedical materials. In: Ratner BD (ed) Progress in biomaterials engineering, vol 6. Elsevier, Amsterdam, p 13
72. Vidrine DW (1982) Photoacoustic Fourier transform infrared spectroscopy of solids and liquids. In: Fourier transform infrared spectroscopy
73. Fries T (1994) Deutscher Verband für Materialprüfung, p 127
74. Sacher E (1988) The determination of the surface tensions of solid films. In: Ratner BD (ed) Progress in biomaterials engineering, vol 6: Surface characterization of biomaterials. Elsevier, Amsterdam, p 53
75. Owens DK, Wendt RC (1969) J Appl Polym Sci 13:1741
76. Pike FD, Thakkar CR Interfacial tension measurement by an improved Wilhelmy technique. In: Kerker M (ed) Colloid and interface science, p 375
77. Wilhelmy LF (1863) Pogg Ann 119:177
78. Matsunaga T, Ikada Y (1981) J Colloid Interface Sci 84:8
79. Li D, Neumann AW (1992) J Colloid Interface Sci 148:190
80. Johnson RE, Dettre R (1969) Wettability and contact angles. In: Surface and colloid science. Wiley, New York, p 85
81. Norde W, Rouwendal E (1990) J Colloid Interface Sci 139:169
82. van Wagenen RA, Andrade JD (1980) J Colloid Interface Sci 76:305
83. van Wagenen RA, Coleman DL, King RN, Triolo P, Brostrom L, Smith LM, Gregonis DE, Andrade JD (1981) J Colloid Interface Sci 84:155
84. Lyklema J (1985) Interfacial electrochemistry of surfaces with biomedical relevance. In: Surface and interfacial aspects of biomedical polymers, p 293
85. Jacobasch HJ, Bobeth W (1975) Das Papier 29:555
86. Harlow E, Lane D (1989)
87. Peters JH, Baumgarten M (1990) Monoklonale Antikörper, 2 Aufl, Berlin, Heidelberg
88. Childs RE, Bardsley WG (1975) Biochem J 145:93

89. Nygren H, Stenberg M (1988) J Biomedical Mat Research 22:1
90. Brauner A, Augthun M, Kaden P (1988) Z Zahnärztl Implantol 4:223
91. Augthun M, Brauner A, Kaden P, Mittermayer C (1988) Z Zahnärztl Implantol 4:228
92. Kirkpatrick CJ, Otterbach T, Anderheiden D, Schiefer J, Richter H, Höcker H, Mittermayer C, Dekker A (1992) Cells & Materials 2:169
93. Courtney JM, Travers M, Douglas JT, Lowe GDO, Forbes CD, Ryan CJ, Bowry SK, Prentice CRM (1986) Assessment of blood compatibility. In: Dawids S, Bantjes A (eds) Blood compatible materials and their testing. Martinus Nijhoff Publishers, Dortrecht, p 129
94. Anderson JM, Kottke-Marchant K (1985) CRC Critical Reviews in Biocompatibility 1:111
95. Ikada Y, Iwata H, Horii F, Matsunaga T, Taniguchi M, Suzuki M, Taki W, Yamagata S, Yonekawa Y, Handa H (1981) J Biomed Mat Res 15:697–718
96. van Wachem PB, Hogt AH, Beugeling T, Feijen J, Bantjes A, Detmers JP, van Aken WG (1987) Biomaterials 8:323
97. Kaelble DH, Moacanin J (1977) Polymer 18:475
98. Coleman DL, Gregonis DE, Andrade JD (1982) J Biomed Mat Res 16:381
99. Bantjes A (1978) Brit Polym J 10:147
100. Klee D, Severich B, Höcker H (1996) Makromol Chem Makromol Symp 103:19
101. Yasuda HK, Yeh YS, Fusselman S (1990) Pure Appl Chem 62:1698
102. Klee D, Höcker H (1995) Spektrum der Wissenschaft 6:90
103. Höcker H, Klee D (1996) Macromol Symp 102:421
104. López GP, Ratner BD, Rapoza RJ, Horbett TA (1993) Macromolecules 26:3247
105. Hoffman AS (1982) Biomaterials 199:1
106. Boenig HV (1982) Plasma science and technology. Carl Hanser, München
107. Kang ET, Neoh KG, Tan KL, Uyama U, Morikawa N, Ikada Y (1992) Macromolecules 25:1959
108. Rose PW, Liston EM (1985) Plastics Engineering 41:41
109. Kang ET, Neoh KG, Tan KL (1992) Macromolecules25:6842
110. Horbett TA (1986) Hydrogels in medicine and pharmacy, 1st edn. CRC Press, New York
111. Mok S, Worsfold DJ, Fouda A, Natsuura T (1994) J Appl Polym Sci 51:193
112. Hsiue GH, Lee SD, Wang CC, Hung-Ishiue M, Chuen-Tsuei Chang P (1994) Biomaterials 15:163
113. Thelen H, Kaufmann R, Klee D, Höcker H (1995) Fresenius J Anal Chem 353:290
114. Thelen H (1996) PhD thesis, Technical University of Aachen
115. Andrade JD, King RN, Gregonis DE, Coleman DL (1979) J Polym Sci Symp 66:313
116. Elwing H, Askendal A, Lundström I (1987) Prog Colloid Polym Sci 74:103
117. Sheu MS, Hoffmann AS, Feijen J (1992) J Adhes Sci Technol 6:995
118. Plüster W, PhD thesis, RWTH Aachen (1998)
119. Cryer D (1983) Biomedical interactions at the endothelium. Elsevier, Amsterdam
120. Potts JR, Campbell ID (1985) Curr Opin Cell Biol 6:648
121. Mc Donagh J (1985) Plasma fibronectin. Academic, New York
122. Dickinson CD, Veerapandian B, Dia XP, Hamlin RC, Xuong NH, Ruoshlahti E, Ely KR (1994) J Mol Biol 236:1079
123. Giroux TA, Cooper SL (1991) J of Appl Poly Sci 43:145
124. Clark DT (1982) Pure Appl Chem 54:415
125. Briggs D, Kendall CR (1982) Int J Adhesion Adhes :13
126. Giga J, Verbist JJ, Josseaux P, Mesmaeker AK (1985) Surf Interface Anal 7:163
127. Wren AG, Phillipps RW, Tolentino LU (1979) J Colloid Interface Sci 70:544
128. Thiele L (1979) Acta Polymerica 30:323
129. Harlow E, Lane D (eds) Antibodies, a laboratory manual. Cold Spring Harbor Laboratory (1989)
130. Klee D, Villari RV, Höcker H (1994) J Mater Sci: Mater Med 5:592
131. Tamada Y, Ikada Y (1993) Polymer 34:2208

132. Mittermayer C, Klee D, Richter H European Patent 0290642,Verfahren zur Besiedlung einer Polymeroberfläche mit menschlichen Gefäßinnenhautzellen
133. Anderheiden D, Brenner O, Klee D, Kaufmann R, Richter HA, Mittermayer C, Höcker H (1991) Angew Makromolekul Chem 185:109
134. Lorenz G, Klee D, Höcker H, Mittermayer C (1995) J Appl Polym Sci 57:391
135. Trevar MD (1980) Immobilized enzymes. Wiley, New York
136. Brandt J, Andersson LO, Porath J (1975) Biochim Biophys Acta 386:196
137. Breuers W, Klee D, Höcker H, Mittermayer C (1991) J of Mater Sci: Mater Med 2:106
138. Anderheiden D (1989) Diplomarbeit am Lehrstuhl für Textilchemie und Makromolekulare Chemie. RWTH Aachen
139. Anderheiden D, Trommler A, Heller B, Klee D, Dekker A, Richter HA, Kirkpatrick CJ, Mittermayer C, Höcker H (1992) Adv Biomaterials 10:199
140. Schroeder K, Klee D, Höcker H, Leute A, Benninghoven A, Mittermayer C (1995) J Appl Polym Sci 58:699
141. Breuers W, Klee D, Plein P, Richter HA, Menges G, Mittermayer C, Höcker H (1987) Kunststoffe 77:1273
142. Mittermayer C, Briolant F, Richter HA, Klee D, Anderheiden D, Breuers W, Diaz-Nogueira E, Füzesil L, Gamero A, Grau M, Kaden P, Varela-Duran J, Lafraya JA, Loevenich H, Maurin N, Ochea JR, Kirkpatrick CJ, Höcker H (1992) European Technology Awards. Proc 4th BRITE/EURAM, Sevilla, p 90
143. Brandt J, Andersson LO, Porath J (1975) Biochim Biophys Acta 386:196
144. Briggs D (1983) Applications of XPS in polymer technology. In: Briggs D, Seah MP (eds) Practical surface analysis by Auger and X-ray photoelectron spectroscopy. Wiley, New York, p 359
145. Dennis AM, Howard RA, Kadish KM, Bear JL, Brace J, Winograd N (1980) Inorg Chim Acta 44:139
146. Schulze PD, Shaffer SL, Hance RL, Utley DL (1983) J Vac Sci Technol 1:97
147. Lorenz G (1995) PhD thesis, Technical University Aachen, Shaker Verlag, Aachen
148. Wahlgren M, Arnebrant T (1991) Tibtech 9:210
149. Wirsching J, Klee D, Höcker H, Dekker A, Mittermayer C (1995) Proc 12th European Conference on Biomaterials, Porto, p 104
150. Schäfer K, Höcker H (1996) Melliand Textilber 77:402
151. Klee D, Lorenz G, Wirsching J, Bienert H, Dekker A, Höcker H, Mittermayer C (1996) Proc 5th World Biomaterials Congress, Toronto, p 562
152. Skechan P, Storbuy R., Scudiero D (1990) J Natl Cancer Inst 82:1197
153. Langefeld S, Numan CJ, von Fischern T, Becker J, Bienert H, Völcker N, Reim M, Schrage NF (1997) Ophthalmic Research 29(S1):78
154. Völker N, Klee D, Höcker H, Langefeld S (1998) J Mater Sci: Mater Med, in press

Editor: Prof. G. Wegner
Received: Dec. 1998

Poly(ε-caprolactone) Blends

G.C. Eastmond

Department of Chemistry, Donnan Laboratories, University of Liverpool,
Liverpool L69 7ZD, UK
e-mail: eastm@liv.ac.uk

Poly(ε-caprolactone) is unusual in being considered compatible with several polymers to produce useful polymer blends. Such blends have been widely investigated both with regard to producing blends with enhanced physical or mechanical properties and to plasticizing rigid polymers with a polymeric plasticizer which will not leach from its substrate. These several materials have potential practical applications. In addition, blends of poly(ε-caprolactone) are probably the most widely investigated series of blends involving one or more crystallisable components and, in this context, provide information which enhances the general understanding of polymer blends.

Keywords. Poly(e-caprolactone), Blend, Phase behaviour, Crystallisation, Glass-transition temperature, Properties

Advances in Polymer Science, Vol.149
© Springer-Verlag Berlin Heidelberg 2000

List of Symbols and Abbreviations

α	coefficient of thermal expansion
α	average distance between junction points of block copolymers in blends
α_0	average distance between junction points of block copolymers in pure copolymer
α_A	difference between coefficients of thermal expansion of liquid and glassy forms of specified component
A, B	unspecified polymers
AN	acrylonitrile
β	compressibility
β	a cooling rate in non-isothermal crystallisation
B	interaction energy density
χ, χ_{AB}	Flory-Huggins interaction parameter between unspecified or specified components
$(\chi_{AB})_{cr}$	critical value of the interaction parameter
c, c_A	concentration of an unspecified or specified polymer in solution
C	a constant (temperature) used in Eq. (27)
CAB	cellulose acetate butyrate
CDA	cellulose diacetate
CTA	cellulose triacetate
CPE	chlorinated polyethylene
CPE-x	chlorinated polyethylene containing x wt % chlorine
δ	phase angle in DMTA

δ, δ_A	solubility parameter of unspecified component or of A
ΔEv	energy of vaporisation of volume V of a substance
d	length scale in spinodal decomposition
D	spherulite diameter
D	domain spacing in block copolymer blend
D_o	domain spacing in pure block copolymer
DCE	1,2-dichloroethane
ΔC_p	change in specific heat at the glass-transition temperature
ΔF^*_m	a parameter used in Eq (26) and defined in text
ΔG_m	free energy of mixing
Δh_{Bu}	heat of fusion of component B
ΔH_f	enthalpy of fusion
ΔH_m	enthalpy of mixing
DMA	dynamic mechanical analysis
DMF	N,N-dimethyl formamide
ΔS_c	combinatorial entropy of mixing
ΔS_m	entropy of mixing
DSC	differential scanning calorimetry
DTA	differential thermal analysis
$[\eta], [\eta]_A$	intrinsic viscosity of a blend or specified polymer
E'	storage modulus in dynamic mechanical analysis
E"	loss modulus in dynamic mechanical analysis
EB	ethylbenzene
$E(t)$	volume fraction of crystallisable component crystallised at some time t
η, η_A	viscosity of unspecified or specified component
η_{sp}	specific viscosity
G_m	a growth rate in crystallisation kinetics
G_0	a constant in crystallisation kinetics
H	hexane
HMTA	hexamethylenetetramine
4HS	4-hydroxystyrene
HV	hydroxyvalerate
I_∞	plateau value of scattering intensity in WAXS experiments
k	a scaling parameter
k	a parameter defined in Eq. (24)
k	a rate coefficient in the Avrami equation, Eq. (26)
K_n	a rate coefficient in the Avrami equation modified for non-isothermal crystallisation
$K(t)$	a cooling function in non-isothermal crystallisation
λ	X-ray wavelength
L	periodicity in ring-banded spherulites
L	long-spacing in SAXS experiments
LCP	liquid-crystalline polymer
LCST	lower-critical-solution temperature

m	a scaling parameter
MA	maleic anhydride
MEK	methylethylketone
$\overline{M}_n$	number-average molecular weight, quoted in g mol^{-1}
$\overline{M}_w$	weight-average molecular weight, quoted in g mol^{-1}
$\overline{M}_w/\overline{M}_n$	polydispersity
v_{Bu}	volume of component B
n	exponent in unmodified and modified Avrami equations
oligo-PS	oligomeric polystyrene
P4HS	poly(4-hydroxystyrene)
P4HS	poly(4-hydroxystyrene-co-styrene) with x mol % 4HS
PB	polybutadiene
PC	bisphenol-A polycarbonate
PCL	poly(ε-caprolactone)
PDTC	poly(2,2-dimethyltrimethylene carbonate)
ψ_B	a parameter defined in Eq. (14)
PHB	poly(3-hydroxybutyrate)
PHBV	poly(3-hydroxybutyrate-co-3-hydroxyvalerate)
PMHS	partially methoxylated poly(4-hydroxystyrene)
PMMA	poly(methyl methacrylate)
PN	propionitrile
PP	n-propyl propionate
PS	polystyrene
PVC	poly(vinyl chloride)
PVME	poly(vinyl methyl ether)
ϕ_A, ϕ_i	volume fractions of components A or I, etc.
$(\phi_A)_{cr}$	critical value of the volume fractions of component A
r_A, r_i	number of units in chains of polymers A or I, etc.
R	the gas constant
σ	lateral surface free energy
σ_e	crystal surface free energy
$(\sigma\sigma_e)^{1/2}$	surface free energy
S	styrene
SAA	styrene-co-allyl alcohol copolymer
SAN	styrene-acrylonitrile copolymer
SAN-n	styrene-acrylonitrile copolymer containing n wt % AN
SAXS	small-angle X-ray scattering
SMA	styrene-maleic anhydride copolymer
SMA-n	styrene-maleic anhydride copolymer with n wt % maleic anhydride
SSEBS	styrene-(ethylene-co-butylene)-styrene triblock copolymer
$\tau_{1/2}$	time taken to reach half plateau value of scattering intensity I
θ	scattering angle
t	a time
$t_{1/2}$	half-time for crystallisation
T	temperature

$<T_a>n$	average thickness of amorphous layers in block copolymer blends
$<T_c>n$	average thickness of crystalline layers in block copolymer blends
T_c	crystallisation temperature
TEOS	tetraethoxysilane
T_g, T_g	glass-transition temperature of unspecified component or a blend
T_{gA}	glass-transition temperature of a specified component
THF	tetrahydrofuran
T_m	crystal melting temperature
T^0_m	equilibrium melting point of the pure crystallisable substance
T^0_{mb}	equilibrium melting point in the blend
TMPC	tetramethyl bisphenol-A polycarbonate
UCST	upper critical solution temperature
V	a total volume
V_{au}, V_{Bu}	molar volumes of the individual repeat units
V_0	a reference volume
V_r	a reference volume
ωa_T	frequency in rheological measurements
w_A	weight fraction of component A
WAXS	wide-angle X-ray scattering
XPCL	poly(ε-caprolactone) with terminal carboxyl groups
X_t	fraction of crystallisable polymer which has crystallised at time t
$X(t)$	fraction of crystallisable polymer which has crystallised at time t in nonisothermal crystallisation

1
Introduction

The prospect of producing new materials with technologically useful properties by mixing different polymers, to form the equivalent of solutions or metal alloys, has long been an attractive goal. Technological aims in studies of polymer blends have been to enhance some useful property, to reduce the cost of some useful attribute or to eliminate some deleterious property of one component. There have also been investigations aimed at generating an understanding of the fundamental nature of polymer mixtures. Unfortunately most pairs of chemically-different polymers are mutually immiscible and mixing usually results in materials which are phase separated and have weak polymer-polymer interfaces. Many polymer blends, therefore, have very poor mechanical properties.

Poly(ε-caprolactone), with chemical structure 1, is unusual in that it has been reported to be miscible with several other polymers and is also considered to produce compatible, although phase-separated, blends with another group of polymers. In this context, miscible implies mutual solubility of polymers on a scale approaching molecular, and compatible implies the existence of useful combinations of properties for possible practical uses.

$$\left[\begin{array}{c} \\ -O-CH_2-CH_2-CH_2-CH_2-CH_2-\overset{\displaystyle O}{\overset{\displaystyle \|}{C}} \\ \end{array} \right]$$

Structure 1

$$\left[\begin{array}{c} \\ -O-\left(CH_2 \right)_n \overset{\displaystyle O}{\overset{\displaystyle \|}{C}} \\ \end{array} \right]$$

Structure 2

Poly(ε-caprolactone) (PCL) consists of sequences of methylene units separated by ester groups and is one member of the general series of aliphatic polyesters 2 which have different methylene to carbonyl ratios, i.e. have different values of n. In these structures there is little steric hindrance to rotation about main-chain bonds and, in consequence, amorphous PCL has a low glass-transition temperature and is soft and rubbery. Regularity in its molecular structure renders PCL highly crystallisable and the crystalline entities reinforce the material and enhance its mechanical properties.

Substances with low glass-transition temperatures are often added to polymers of high glass-transition temperature in order to render them more flexible and to reduce brittleness; such substances, which must be miscible with the polymer having a high glass-transition temperature, are referred to as plasticizers. Conventional, low-molecular-weight plasticizers include esters, such as phthalates, or tricresyl phosphate. A major disadvantages of conventional low-molecular-weight plasticizers is that they may leach out of the polymer over extended periods of time with a consequent loss of plasticization. The polymers then become rigid and susceptible to cracking and, possibly, shrinkage. Polymeric plasticizers are therefore potentially attractive as they are unlikely to leach out of the polymeric matrix on any reasonable timescale. PCL, being miscible with several polymers, is therefore interesting as a polymeric plasticizer. Early publications in the field of PCL blends recognised this potential [1, 2].

This review is not intended to provide a catalogue of possible applications of PCL blends but describes investigations into the properties of such blends and develops an appreciation of the nature and properties of them. In this sense, immiscible blends are of little interest because they consist, to a first approximation, of laminates of PCL with the second component or of dispersions of, normally, the minor phase within a matrix of the major phase, according to the method of fabrication. This review concentrates on miscible and, more commonly, partially miscible systems in which polymer properties and morphologies vary with blend composition, thermal and processing history, etc.

However, for illustrative purposes, we refer to two different applications of immiscible blends involving PCL. In one, solution blending of PCL with polycaprolactam (nylon 6) from solution in 1,1,1,3,3,3-hexafluoro-2-propanol (hexafluoroisopropanol) is said to improve the impact resistance of nylon 6 [3]. In the second example, blends of PCL and polyisoprene give blends which are both photo- and biodegradable; one component is said to degrade by each mechanism, i.e. PCL is biodegradable and polyisoprene is photodegradable [4].

At the other extreme, polymeric plasticization requires molecular miscibility and examples of such plasticized systems which have been claimed are quoted in Sect. 6.

Compounding polymer blends often involves the use of high processing temperatures in order to induce sufficient molecular mobility into the component which softens at the higher temperature. All esters are susceptible to transesterification reactions as is PCL when mixed with other polyesters or polycarbonates under such conditions. These processes provide added complications to blending processes by producing copolymers in situ. A recent patent refers to the use of such processes in blending PCL with polyesters or polycarbonate to produce antielectrostatic compounds [5]. To date, most studies of transesterification reactions have been made on aromatic polyesters but there have recently been a few additional studies on PCL blends with other polyesters; these are discussed in Sect. 12 on blends with polyesters.

Block copolymers, formed by transesterification of blend components or otherwise, will act as interfacial agents in blends of immiscible polymers which are identical to or compatible with the components of the block copolymer. A recent patent application for a system of this type refers to the use of triblock copolymers of polymer A-polysiloxane-polymer B as interfacial agents in blends of polymers A and B which have different polarity or hydrophilicity [6]. In this case the intermediate polysiloxane blocks, strongly immiscible with either polymer A or polymer B, helps to drive the copolymer to the A-B interface and reduce the possibility of dissolution of the copolymer in either blend phase. In the example quoted polymer B is PCL (or a non-polar polyether) and polymer A is a polar polyether or other polar polymer.

Block or graft copolymers (graft copolymers have the same influence as interfacial agents) may also be prepared in situ, not by transesterification but by reactive blending. An example of a system involving PCL has been given by Cavallaro et al. [7, 8]. Poly(β-hydroxybutyrate-*co*-β-hydroxyvalerate) and PCL were blended in the presence of a peroxide. Decomposition of the peroxide generated radicals which could abstract a hydrogen atom from a polymer, probably the poly(β-hydroxybutyrate-*co*-β-hydroxyvalerate), to generate a macroradical and then, through chain-transfer reactions, PCL macroradicals could be formed which could undergo radical-radical recombination reactions and generate graft copolymers. Spectroscopic, DSC and thermogravimetric data were quoted to support the formation of copolymer in this system.

Thus, there are many complexities in the study of PCL blends. The various systems studied often lead to non-equilibrium structures and properties which may, however, have practical utility. This complexity is exacerbated by the fact that the properties of PCL blends might vary with the origin of the component polymers. It has been noted that some commercial samples of PCL appear to contain a nucleating agent for PCL crystallisation [9]. Only one study has commented on this feature but it inevitably follows that the nature and properties of blends might depend on the source of the polymers used and will depend on whether or not such agents are present.

2
Poly(ε-caprolactone)

Poly(ε-caprolactone) 1 is a simple, linear, aliphatic polyester formed by the ring-opening addition polymerisation of ε-caprolactone, normally initiated by an alcohol or diol in the presence or absence of a catalyst. The polymer has a regular structure and is crystallisable. PCL crystallises to about 50% in the form of spherulites [10].

Within the crystalline fibrils (see Sect. 4) the chains crystallise in lamellae as extended chains in the planar zig-zag conformation [11, 12], akin to that in crys-

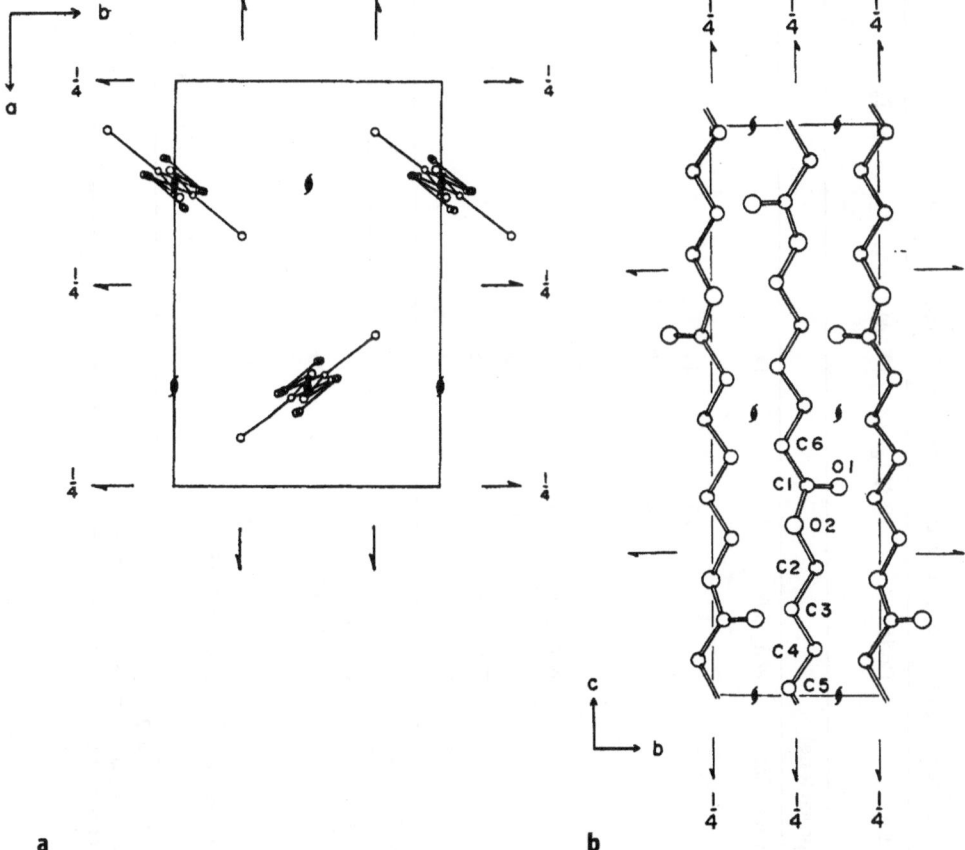

Fig. 1a,b. Projections of the crystal structure of poly(ε-caprolactone) viewed along: **a** the a-axis; **b** the c-axis

Table 1. Characteristics of some commercial samples of PCL and poly(vinyl chloride) used in studies of PCL blends

Manufacturer	Code	$10^{-3}\overline{M}_n$	$10^{-3}\overline{M}_w$	T_m °C	T_g °C	Viscosity/dl g⁻¹ Inherent	Intrinsic	references: values Determined in	Quoted in
Poly(ε-caprolactone)									
Union Carbide Corp.	PCL-700	13	14	61	-60		0.416ᵃ	[14]	[15]
		23.2	45						[16]
		viscosity	ave. ~40						[9]
		15.5	40.4					[18]	[17]
			40						[19, 20]
									[21]
	PCL-300	10	15						[21]
		10							[22]
	Tone P787		~80						[23]
	Tone P767		~30						[23]
Poly(vinyl chloride)									
Union Carbide Corp.	QYTQ					1.0ᵇ			[2]
	QYTQ-387	35	70					[14]	[15, 16]
			78				1.005ᶜ		[16]
		38							[9]
Kema Nord AB, Sweden	Pevikon S-655		74		95 (1 Hz)				[17]

ᵃ Determined in THF at 25° C
ᵇ Determined in cyclohexanone at a concentration of 0.2 wt %
ᶜ Determined in THF at 25° C, converted to a molecular weight of 86,400 using relation in [24]

talline polyethylene. The unit cell is orthorhombic and perpendicular to the chain direction; unit cell dimensions are very similar to those of polyethylene and the planar zig-zags are at similar angles to the α-axis, with alternate layers in opposite directions (Fig. 1a). Detailed analyses of fibre diffraction patterns differ slightly [11, 12]. Both groups agree that the sequence of CH_2 units is in a planar zig-zag conformation. Chatani et al. conclude that the ester groups are slightly twisted from the plane of the planar zig-zags, which results in a slight shortening of the chain dimensions from those its fully-extended form [12]. In the main-chain direction (c-axis) the ester groups are in alternate directions, (Fig. 1b) as would conform to a chain-folded structure and as known for nylon-6, and carbonyl groups on adjacent chains are shifted relative to their neighbours. The two groups of workers determined very similar unit cell dimensions:

Bittiger et al. [11] found a=7.496, b=4.974, c=17.297 Å.

Chatani et al. [12] found a=7.47, b=4.98, c=17.05 Å.

The crystalline polymer has a melting point of about 63 °C. The wholly amorphous polymer has a glass-transition temperature (T_g) of about –70 °C [2] and is therefore rubbery; because of the strong tendency of PCL to crystallise, even when quenched rapidly to low temperature from the melt, it is difficult to determine an unambiguous value for T_g for the wholly amorphous polymer. In the partially crystalline state the high-molecular-weight polymer is tough and flexible but low-molecular-weight polymer is brittle. The T_g of the amorphous component in the partially crystalline polymer is about –60 °C [2, 13].

Several studies of PCL blends have made use of commercially available polymers. The characteristics of the major samples of PCL and also of poly(vinyl chloride) are given in Table 1; not all authors quote consistent information for nominally the same polymer and use of incorrect molecular weights could affect the conclusions from discussions of properties, especially of miscibility. Properties of the samples are detailed at relevant points in the text [2, 9, 14–24].

3
Thermodynamics of Mixing

It is the thermodynamics of mixing which ultimately determines whether or not a blend is miscible or immiscible at equilibrium. The phase diagram, which is determined by the thermodynamics of mixing, defines the conditions under which homogeneous mixtures are stable, metastable or unstable and will predict whether a blend will become miscible or immiscible as experimental parameters are changed.

A full discussion of the thermodynamics of polymer blends is beyond the scope of this review and only some essential, and frequently discussed, aspects are presented here in order to provide the reader with some principles of the thermodynamics of polymer blends.

The treatment presented here devolves from the thermodynamics of dilute polymer solutions developed by Huggins [25, 26] and by Flory [27–29], now known as the Flory-Huggins theory. This theory is based on a simple lattice model in which a solvent molecule or a segment of a polymer chain can occupy a lattice site; polymer segments within the same polymer molecule occupy a contiguous series of lattice sites. The treatment was extended to polymer-polymer mixtures, i.e. to polymer blends by Scott [30] and by Tompa [31]. While the simple lattice model fails to provide a full description and understanding of many real polymer blends it does suffice to demonstrate the essential origins of the normal immiscibility of chemically-different polymers and the need for specific interactions to promote miscibility. A fuller understanding of the thermodynamics of polymer blends requires knowledge of equation-of-state theories. These theories can overcome the often incorrect temperature-dependence of phase behaviour embodied in the simple lattice model. Such theories are not discussed here and the interested reader is referred to other texts and to the original references cited therein [32–36].

3.1
Amorphous Blends

True miscibility of chemically-different polymers implies favourable thermodynamics of mixing; a major (but insufficient) criterion for mutual miscibility is that the free energy of mixing (ΔG_m) must be negative (Eq. 1). In small molecule systems the major driving force for miscibility is a high entropy of mixing (ΔS_m); $T\Delta S_m$ is large. The enthalpy of mixing (ΔH_m) is usually positive and opposes mixing:

$$\Delta G_m = \Delta H_m - T\Delta S_m \tag{1}$$

The ideal entropy of mixing can be estimated by counting the number of ways in which two sets of different polymer chains can be arranged on a three-dimensional lattice and simultaneously fill space; this simple lattice model is the basis of the Flory-Huggins theory of polymer solutions and its extension to polymer blends. The result, assuming that all lattice sites are of equal volume and can be occupied by one chain segment of either polymer, is the combinatorial entropy of mixing (ΔS_c) which, for total volume V, is given by

$$\Delta S_c = -\frac{RV}{V_o}\left\{\frac{\phi_A}{r_A}\ln\phi_A + \frac{\phi_B}{r_B}\ln\phi_B\right\} \tag{2}$$

where ϕ_i and r_i are the volume fractions and number of segments per chain of polymer i, R is the gas constant and V_0 is a reference volume equal to that of a lattice site. It is sometimes assumed that the number of segments per chain is the number of monomer units per chain, or the degree of polymerisation, but this is often inappropriate. In mixtures of polymers ΔS_m or ΔS_c decreases with increasing molecular weights, or number of segments per chain, of the components. For high-molecular-weight polymers ($r_i > 100$, approximately) ΔS_m is negligible.

The same positive ΔH_m, which arises from interactions between different species, usually prevails in polymer mixtures and most polymer pairs are mutually immiscible. The strengths of the interactions between components is usually expressed in terms of an interaction parameter χ_{AB} (Eq. 3), which originates from Flory-Huggins theory of polymer solutions. The enthalpy of mixing, or interaction energy term, arises in a van Laar type expression of heats of mixing and is of the form

$$\Delta H_m = RT\chi_{AB}\phi_A\phi_B \tag{3}$$

The interaction parameter χ_{AB} is a measure of the energy involved in creating a contact between unlike components previously in contact with like components. The interaction energy is determined by χ_{AB} and the volume fractions of the components.

Although Flory-Huggins theory does not give an accurate description of polymer blends, it identifies the origins of polymer-polymer immiscibility and usually forms the basis of discussions; interaction parameters are usually calculated assuming its applicability. For a more accurate description of the thermodynamics of mixing it is necessary to use a more complete equation-of-state theory [37, 38].

For polymer-polymer miscibility it is normally necessary that χ_{AB} is negative. In principle values of χ_{AB} should be independent of composition but experimental evidence often indicates a composition dependence of χ_{AB} which reflects the inadequacy of the simple theory.

Miscibility in blends of high-molecular-weight polymers, therefore, normally requires ΔH_m to be negative and this implies the existence of favourable, specific interactions between the different polymers, such as the formation of hydrogen bonds or acid-base, charge transfer or other interactions. Interactions between small molecules, corresponding to negative values of ΔH_m, could, in principle, cause an increase in contacts between unlike components, greater than that predicted by random mixing, and modify ΔS_m. Positive values of ΔH_m could reduce the number of contacts between unlike components. It is usually considered that such interactions do not modify the entropy of mixing in mobile (small molecule) systems. Solutions with sufficient thermal energy to overcome such effects are known as regular solutions and ΔS_m approximates to its ideal value. The same assumption is not valid in polymer blends where interactions between units in the different components may cause sufficient inter-chain associations to modify entropies of mixing in some miscible systems.

Because polymer-polymer miscibility depends on a balance between ΔH_m and ΔS_m and because ΔS_m is molecular weight dependent, any statement as to miscibility, partial miscibility or immiscibility should be qualified by the molecular weights of the polymers involved. A system which is partially miscible for one pair of molecular weights may become miscible for mixtures of lower molecular weight species or immiscible if the molecular weight of one or both components is increased.

The overall equation for ΔG_m therefore, from Flory-Huggins theory, becomes, in one of its more common forms

$$\Delta G_m = \left(\frac{RTV}{V_r}\right)\left\{\frac{\phi_A}{r_A}\ln\phi_A + \frac{\phi_B}{r_B}\ln\phi_B + \chi_{AB}\phi_A\phi_B\right\} \tag{4}$$

Even if values of ΔG_m are negative, mixtures are not necessarily miscible at all compositions. It is possible that such mixtures can achieve a lower total free energy by undergoing phase separation to give two phases of greater stability.

Figure 2 shows plots of the term in braces on the right-hand-side of Eq. (4) as functions of volume fractions of polymer A for binary blends of polymers A and B of unequal degrees of polymerisation; plots for mixtures of polymers of equal degrees of polymerisation are symmetrical. These plots show changes in the relative free energies of mixing for different positive values of χ_{AB}. Curve a ($\chi_{AB}=0$) represents the combinatorial entropy of mixing only which is small and favourable for mixing at all compositions; the curve is concave upwards for all values of ϕ_{AB}. Negative values of χ_{AB} give similar curves with larger negative values of ΔG_m.

Curve d in Fig. 2 depicts a situation with a modest positive value of χ_{AB} but sufficiently large as to make ΔG_m positive at some compositions. Obviously, mixing under such conditions is not favourable.

Curve c represents an intermediate situation in which χ_{AB} is sufficiently small that ΔG_m is always negative but the curve has points of inflexion. Despite values of ΔG_m being negative, mixtures with compositions between those at which a

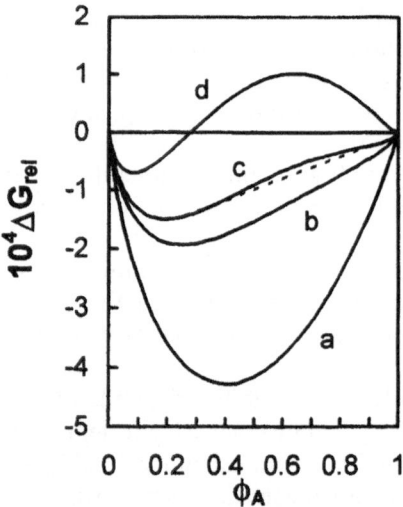

Fig. 2. Variations in relative free energies of mixing in a binary blend of polymers A and B with degrees of polymerization of, respectively, with volume fraction of polymer A for different values of the interaction parameter: (a) 0, (b) 0.00105, (c) 0.0013, (d) 0.0020

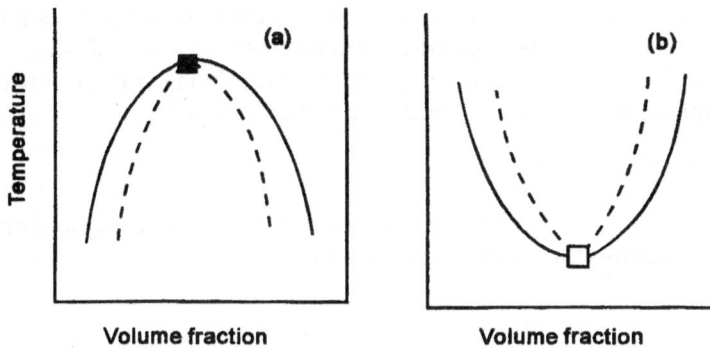

Fig. 3a,b. Schematic representation of phase diagrams showing; **a** upper critical solution temperature (UCST); **b** lower critical solution temperature (LCST) behaviour. Binodals (—) and spinodals (----) are shown together with UCST(≠) and LCST( ) points

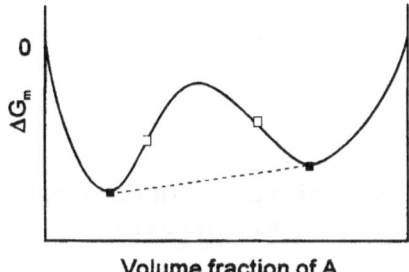

Fig. 4. Schematic representation of the variation in free energy of mixing with volume fraction of polymer A in a binary mixtures of polymers A and B: *filled squares* define compositions corresponding to the binodal and *open squares* the spinodal of the phase diagram

common tangent (broken line) meets curve c can achieve a lower total (more negative) free energy state by phase separating into phases defined by the points of contact of the common tangent with the curve. These points of contact define the binodal of the phase diagram, shown in Fig. 3, and represent the boundaries between homogeneous and phase-separated mixtures at equilibrium; compositions outside the binodal are homogeneous and those inside are heterogeneous at equilibrium.

This situation is seen more clearly in Fig. 4, showing a schematic curve in which the same effect is depicted more strongly. For curves such as c in Fig. 2 and the curve in Fig. 4 there are compositions between which the curve is concave downwards, i.e. between points of inflexion. These points of inflexion (open squares in Fig. 4) define the spinodal of the phase diagram as shown in Fig. 3. The further significance of these points is discussed later. Between the points of inflexion $\partial^2\Delta G/\partial\phi^2<0$; $\partial^2\Delta G/\partial\phi^2=0$ at the spinodal. Filled squares in Fig. 4 are the compositions on the binodal shown in Fig. 3.

If χ_{AB} is decreased to the value applicable to curve b in Fig. 2 a situation is reached where mixtures become miscible at all compositions. This situation corresponds to the critical points of the phase diagrams in Fig. 3 and where the binodal and spinodal compositions merge. At the critical point

$$\partial^2 \Delta G / \partial \phi^2 = \partial^3 \Delta G_m / \partial \phi^3 = 0$$

Thus miscibility throughout the composition range, according to Flory-Huggins theory, requires the further condition that

$$\frac{\partial^2 \Delta G_m}{\partial \phi^2} > 0 \tag{5}$$

at all compositions.

The value of χ_{AB} corresponding to the critical point changes as ΔS_m varies with the degrees of polymerisation of the component polymers and, in general, the critical value is

$$\left(\chi_{AB} \right)_{cr} = \frac{1}{2} \left\{ \left(\frac{1}{r_A} \right)^{1/2} + \left(\frac{1}{r_B} \right)^{1/2} \right\} \tag{6}$$

In addition, the volume fraction at the critical point in Fig. 3, where the binodal and spinodal compositions merge, is given by

$$\left(\phi_A \right)_{cr} = \frac{r_A^{1/2}}{r_A^{1/2} + r_B^{1/2}} \tag{7}$$

For high-molecular-weight polymers the value of $(\chi_{AB})_{cr}$ is very small; values of r_A and r_B in Eqs. (4) and (6) are large. In practice, values of χ_{AB} for apolar polymer mixtures, i.e. for contact between apolar units in chains, are positive and usually exceed the critical value. Therefore, for polymer-polymer miscibility there is a necessity for specific, attractive interactions between chain segments which will render ΔH_m negative.

Enthalpies of mixing can be determined, in principle, calorimetrically but for polymer-polymer mixtures the inherent immiscibility of the components and lack of contacts between them means that these parameters cannot normally be determined directly. It is possible to estimate heats of mixing from mixtures of oligomeric analogues, which are often miscible by virtue of their greater entropy of mixing. The interaction energy density B, related to χ_{AB}, can then be determined from

$$\Delta H_m = B \phi_A \phi_B \tag{8}$$

However such values should be treated with caution because the values so determined may be subject to errors from a series of origins. For example, end groups of oligomers are not identical to main-chain units, they may have different interaction energies with the main-chain units of the second component and, in large concentration, may contribute significantly to the overall enthalpy of mixing. In addition, end groups increase the free volume of the system and hence the free energy of mixing. Also interaction parameters estimated from phase behaviour may include entropic effects because of failings of the concept of regular solutions used in the simple Flory-Huggins approach.

The most important defect in simple theory, from a practical point of view, is its failure to describe the temperature dependence of polymer-polymer miscibility with temperature. Simple theory predicts miscibility at high temperature, irrespective of the value of ΔH_m, since $T\Delta S_m$ inevitably dominates at sufficiently high temperature. This behaviour is characterised by a phase diagram of the type shown schematically in Fig. 3a and is typical of small molecule systems. Such systems are characterised by an upper-critical-solution temperature (UCST). In contrast, polymer-polymer systems (if not immiscible at all accessible temperatures) are usually characterised by a lower-critical-solution temperature (LCST), Fig. 3b, and are more likely to be miscible at low temperatures.

Blends which are miscible at low temperature may undergo phase separation on heating if the temperature to which they are heated exceeds the LCST of the phase diagram and if the compositions fall within the binodal. The failure of simple theory in this regard has its origins in defects in the simple lattice model. The lattice model assumes a lattice site, of some specific size, can be occupied by a unit of either component. In reality, mismatches in sizes of lattice sites and units can lead to an uneven distribution of free volume which can result in LCST behaviour. A second cause of LCST behaviour can arise from specific (exothermic) interactions, such as hydrogen-bonds, between the components. Theoretical calculations using equations of state theory can predict LCST behaviour [39].

In Fig. 3 the binodal curves represent the equilibrium phase boundaries and determine the compositions of the phases in equilibrium at any temperature. The spinodals represent the boundaries between totally unstable (inside the spinodal) and metastable (between the spinodal and binodal) homogeneous phases if created. The curves in Fig. 3 represent the variations in composition of the points in Fig. 4 with temperature. Phases with compositions between the binodal and spinodal are metastable and nucleation is required to initiate phase separation. In contrast, compositions within the spinodal are totally unstable and will inevitably undergo phase separation if there is sufficient mobility to permit the necessary molecular motions. The mechanisms by which phase separation occurs in those two regions are different and give rise to different morphologies in the phase-separated materials. Thus, the curves define phase behaviour at equilibrium and also define the limits of operation of mechanisms of phase separation, when appropriate.

Whether or not the equilibrium phase behaviour of a particular blend is achieved in practice, i.e. whether or not phases corresponding to the filled squares in Fig. 4 are formed, depends on a number of factors. Neither phase separation from homogeneous mixtures nor dissolution of segregated systems can occur in rigid mixtures. Changes in phase behaviour require molecular mobility which, in turn, require the polymer-polymer system to be above its glass-transition temperature or to be plasticized by solvent or other additive. Details may also be complicated by crystallinity in one or more components. Thus, several factors determine whether or not the details of the phase boundary can be determined.

Many polymer-polymer mixtures are totally immiscible at all accessible temperatures. However, mixtures of high-molecular-weight polymers usually exhibit LCST behaviour (at least in principle). Thus, mixtures which are said to be miscible may phase separate on heating if the samples reach the binodal before decomposition sets in. Phase-separated systems of this type may or may not homogenise on cooling according to the relative rates of the processes involved. Polymer mixtures which exhibit UCST behaviour usually involve oligomeric materials when the factors which control dissolution of low-molecular-weight substances (especially the entropy of mixing at high temperatures) dominate.

It is beyond the scope of this article to discuss, in detail, the mechanisms and conditions of phase separation processes and the reader is referred to standard texts for such accounts [38].

3.2
Crystalline Blends

The preceding remarks relate to amorphous polymer mixtures. Not all polymers or their blends are totally amorphous and PCL blends provide a major set of examples of semi-crystalline, as well as amorphous, blends, often in the same system. It is therefore necessary to consider some thermodynamic consequences of crystallinity in blends.

In general a crystalline substance in contact with an amorphous liquid phase in which it has some solubility will exhibit a lower melting point than if in contact with its own pure liquid phase. That is, the substance experiences a melting point depression. Melting point depressions are general and in small molecule systems a major factor in their origin is the increase in entropy gained when crystalline material is transferred to the liquid phase in the presence of a solute, relative to that when the substance melts in contact with its own pure liquid phase. In polymer-polymer systems, however, the entropy factor is negligible and enthalpy changes associated with changes in the numbers of interactions between like and unlike species dominate.

The theory of melting point depressions applied to such polymer-diluent systems was first developed by Flory. For polymer blends discussion is now usually based on the subsequent work of Nishi and Wang [40] who derived the expression

$$\frac{1}{T_{mb}^0} - \frac{1}{T_m^0} = -\frac{RV_{Au}}{\Delta H_m^0 V_{Bu}} \left[\frac{\ln\phi_A}{r_A} + \phi_B \left(\frac{1}{r_A} - \frac{1}{r_B} \right) + \chi_{AB}\phi_B^2 \right] \tag{9}$$

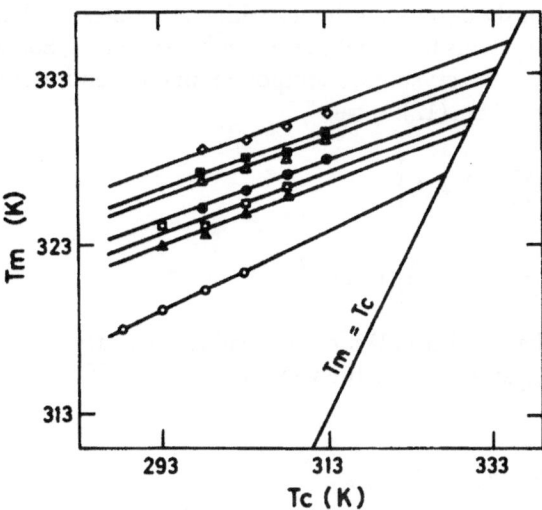

Fig. 5. Variation in melting temperature, T_m, with crystallisation temperature, T_c, for a series of blends of PCL with different proportions of poly(4-hydroxystyrene), illustrating the application of the Hoffman-Weeks procedure; taken from [42]

where ΔH_m^0 is the enthalpy of fusion per mole of repeat unit and T_m^0 is the equilibrium melting point of the pure crystallisable substance, identified as B, and T_{mb}^0 the equilibrium melting point in the blend. The melting point depressions can then be used to estimate values of interaction parameters. For mixtures of two high molecular weight polymers Eq. (9) reduces to Eq. (10):

$$\frac{1}{T_{mb}^0} - \frac{1}{T_m^0} = -\frac{R V_{Au}}{\Delta H_m^0 V_{Bu}} \chi_{AB} \phi_B^2 \tag{10}$$

(V_{Au}, V_{Bu} are the molar volumes of the individual repeat units).

Experimentally observed melting points are usually not the equilibrium melting points. Frequently, in polymeric systems, observed melting points in pure substances also fall well below the equilibrium values. Reduced melting points are often a result of small crystallite size, i.e. lamellar thicknesses, and the consequent high surface area (energy) or of defects in the crystalline regions. Hoffman and Weeks [41] considered the problem of identifying equilibrium melting points in polymeric systems. They proposed the determination of experimental melting points T_m for a series of crystallisation temperatures T_c and extrapolation of a plot of T_m as a function of T_c to its intersection with the line $T_m=T_c$; the intersection is taken to be the equilibrium melting point. This approach has been exemplified in blends of PCL by Lezcano et al. [42] (Fig. 5), amongst others cited in the relevant sections.

Shah et al. applied an extension of the treatment for melting point depression developed for binary blends to ternary blends involving PCL; see Sect. 20 [43].

They used previously developed theory which attributes the heat of mixing in multicomponent systems to a combination of binary terms, Eq. (11), from which the melting point of the crystalline component in the blend, assumed to be component B, is given by Eq. (12):

$$\Delta H_m = V \left\{ \sum_i \sum_{j \neq i} B_{ij} \phi_i \phi_j \right\} \tag{11}$$

$$T_m = T_m^0 \left\{ 1 + B \left(v_{Bu} / \Delta h_{Bu} \right) \right\} \left(1 - \psi_B \right)^2 \tag{12}$$

In Eq. (12) B is given by Eq. (13) for which values of ψ_i are given by Eq. (14) and ΔB by Eq. (15). v_{Bu} and Δh_{Bu} are the volume and heat of fusion of the crystalline component B:

$$B = B_{AB} \psi_A^2 + B_{BC} \psi_B^2 + \Delta B \psi_A \psi_B \tag{13}$$

$$\psi_i = \phi_i / (1 - \phi_i) \tag{14}$$

$$\Delta B = B_{AB} + B_{BC} - B_{AC} \tag{15}$$

Thus, from melting point depressions, if two binary interaction parameters are known, the third binary interaction parameter can be determined.

3.3
Solubility Parameters

Several workers have used solubility parameters, promulgated by Hildebrand and Scott [44], to predict whether or not mixtures of polymers will be miscible. The solubility parameter δ is defined as

$$\delta = (\Delta E_v / V)^{1/2} \tag{16}$$

where ΔE_v is the energy of vaporisation of volume V of a substance and δ^2 is the cohesive energy density of a substance and is a direct measure of the strength of intermolecular interactions. The solubility parameter varies with temperature according to Eq. (17):

$$\delta \cong \frac{T\alpha}{\beta} \tag{17}$$

where α and β are the coefficients of thermal expansion and the compressibility, respectively. If δ_A^2 and δ_B^2 are measures of intermolecular interactions between A species and B species, respectively, then interchanging A and B units to generate AB interactions, where previously only AA and BB interactions existed, and assuming the geometric mean hypothesis, the change in interaction energy is related to $(\delta_A - \delta_B)^2$ where $\delta_A \delta_B$ is a measure of the interaction between A and B units. Thus, the solubility parameter can be related to the enthalpy of mixing by the relation

$$\Delta H_m = V(\delta_A - \delta_B)^2 \phi_A \phi_B \tag{18}$$

which is often used in place of Eq. (8) where values of χ_{AB} are unknown; formation of mixtures is close to athermal when $\delta_A \approx \delta_B$.

This solubility approach has been widely used to predict polymer solubility and swellability when it is usually considered that polymers are preferentially soluble in or swellable by substances (usually solvents) with similar solubility parameters. Closely similar values are said to indicate miscibility and significantly different values to indicate immiscibility. Where values of solubility parameters are unknown they can be estimated by group contribution methods. In performing such calculations it is important to consistently use parameters from a single data set, for example Small's parameters [45] or Hoy's parameters [46]. However, it should be noted that, according to Eq. (18), ΔH_m cannot be negative; $(\delta_A - \delta_B)^2$ must be positive. Thus the solubility parameter approach does not take into account specific interactions, such as hydrogen-bonding, which favour miscibility and lead to negative values of ΔH_m. In this sense the application of solubility parameters in polymer blends is limited. Their use in blends of PVC with polyesters, such as PCL, has been denigrated by Aubin and Prud'homme [47].

3.4
Inverse Gas Chromatography

Inverse gas chromatography is a technique now frequently used to estimate interaction parameters in polymer blends. Substances are separated by affinity chromatography in consequence of their different intermolecular interactions between the mobile and stationary phases. In normal gas chromatography unknown species are separated and/or characterised by their retention times, or retention volumes, when injected into a mobile gas stream, under standard conditions, and interact with a known stationary phase in the chromatography column. In inverse gas chromatography, substances (probe molecules) with known characteristics are injected into a gas stream and interact with a polymer, which requires characterisation, in the stationary phase when interactions result in retention volumes from which interaction parameters can be estimated [33, 48, 49]. Polymers, as the stationary phases, are used as fluids on inert supports at temperatures above their glass-transition or crystal-melting temperatures. This process has been developed to determine interaction parameters between polymers and small molecules, which might be representative of segments of a second polymer [50], and has also been applied to small molecules and polymer mixtures, including blends of PCL [18].

4
Crystal Morphologies

PCL is a highly crystallisable polymer which can crystallise in the pure state or in blends. The crystallisation of PCL and its resulting morphology in the crystalline state accord with classical patterns of polymer crystallisation.

Crystallisable polymers will crystallise at temperatures (crystallisation temperatures T_c) between the crystal melting temperature (T_m) and the glass transition temperature (T_g), i.e.

$$T_g < T_c < T_m$$

An obvious requirement is that $T_c < T_m$ and the thermodynamic driving force for polymer crystallisation increases as T_c decreases. At higher temperatures, as T_c decreases, the increased driving force increases the rate of crystallisation. However, at lower temperatures, as T_c approaches T_g, the viscosity of the system increases, rates of molecular motion decrease and the rate of crystallisation decreases to zero. The overall rate of crystallisation is characterised by a bell-shaped curve with low rates of crystallisation close to T_m and T_g and optimal rates in between.

Crystallisation of polymers such as PCL, which crystallise to give spherulitic structures, starts from a nucleus which subdivides at the growth surface to generate a series of very thin (typically 10 nm thick) crystalline lamellae. The lamellae continue to grow and sub-divide to establish the spherically-symmetric structures (spherulites) which consist of a series of crystalline fibrils, bundles of lamellar crystals, extending from the nucleus in all directions, with a constant

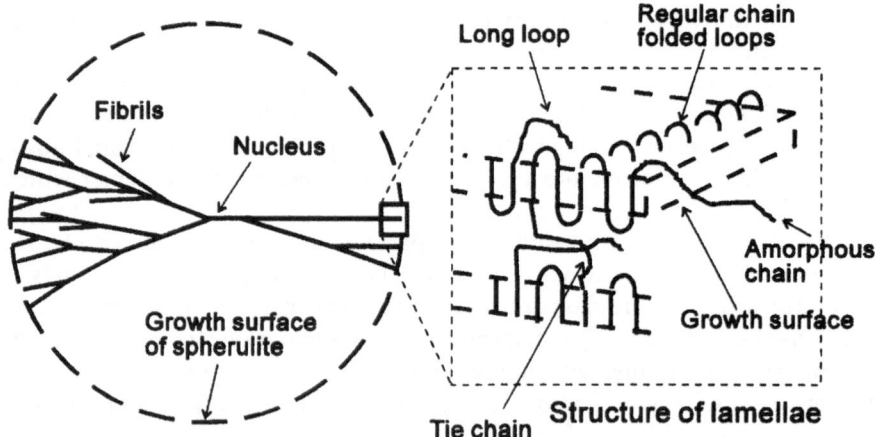

Fig. 6. Schematic representation of the internal structure of a spherulite in a partially crystalline polymer, showing the arrangement of the crystalline fibrils, and of the internal structure of the fibrils showing the arrangement of polymer chains in the lamellae and several local structural features as described in the text

density of fibrils at the outer envelope of the spherulite (Fig. 6). The polymer chains have a characteristic alignment with respect to the fibril length, i.e. to the growth direction (as depicted in Fig. 6), usually perpendicular to the growth direction which is away from the centre of the spherulite. Chains emerging from a lamellar surface may re-enter through regular chain folds or through irregular loops or may enter other lamellae as 'tie chains'. When observed in thin section (i.e. when grown in thin films) by polarising microscopy the spherulites exhibit a characteristic Maltese Cross pattern; with electron microscopy the fibrils may be observed. In some cases the growing fibrils (and their constituent lamellae) twist in a helicoidal manner to form what are known as ring-banded spherulites which, in the polarising microscope, are seen as a series of concentric rings centred on the nucleus (Fig. 7b–d). Both of these features are seen in Fig. 7 [51].

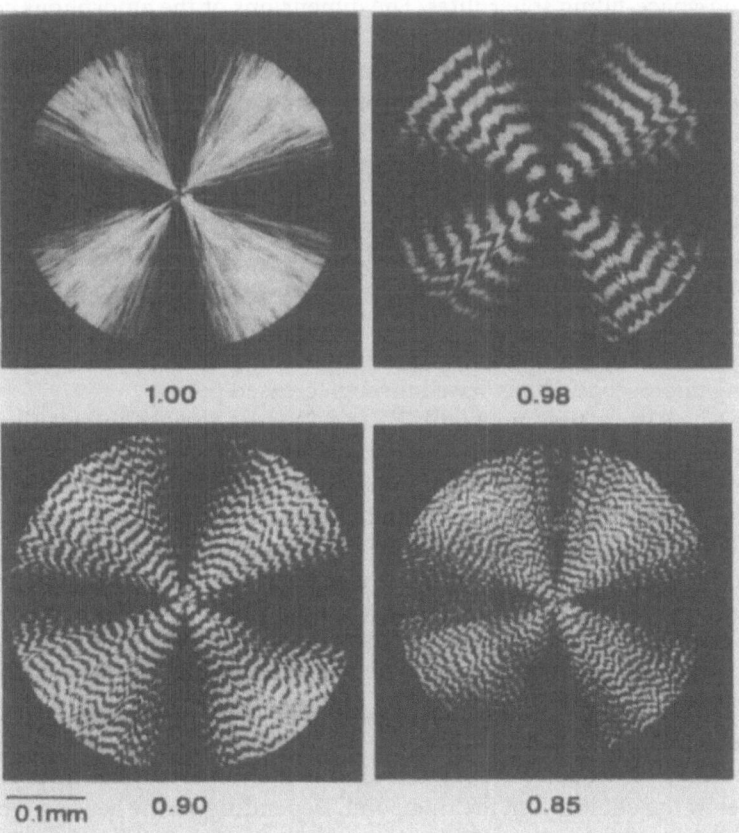

Fig. 7. Optical micrographs (under crossed polars) of PCL spherulites in pure PCL and blends of PCL with PVC; numerals define the weight fraction of PCL in the blends. The micrographs show normal spherulites in pure PCL and ring-banded spherulites in the blends; the periodicity in the blends increases with PCL content; taken from [51]

In polymers with low crystallinity the spherulites, which normally contain the whole of the crystalline material, do not fill space and there is amorphous material between the spherulites. In highly crystalline material the spherulites do fill space and the growing spherulites impinge on each other; they do not interpenetrate but growth stops where they impinge and they adopt polyhedral forms. Nevertheless, no polymer, even highly crystalline polymers with space-filling spherulites, is 100% crystalline. In systems with space-filling spherulites the amorphous material is inevitably located within the spherulites. That is, amorphous and crystalline components coexist and, on heating, samples exhibit both a glass-transition temperature, for the amorphous component, and a melting point, for the crystalline component. This behaviour is characteristic of crystalline polymers and distinct from the behaviour of small-molecule materials. The amorphous polymer is contained between the crystalline lamellae and this polymer exhibits a T_g as does any other amorphous polymer, including that between non-space-filling spherulites. The dimensions of the amorphous component (thickness between fibrils) is small. Polymer chains in the amorphous regions within the spherulites may form links between crystalline lamellae (as tie chains) or exist as other lengths of amorphous chain. Several of these several features are depicted in the right-hand portion of the schematic Fig. 6.

Under constant-growth conditions, e.g. in pure polymer at constant temperature, the rate of growth of the fibrils is constant. Spherulites, therefore, have a constant rate of radical growth. Growth continues until all crystallisable material is consumed or until the spherulites impinge on each other. Where there is insufficient crystallisable material, or when the temperature is decreased too far during crystallisation, the spherulites may not fill space and amorphous polymer may exist between them; the amorphous component appears dark in the polarising microscope when viewed through crossed polars.

In the Nishi-Wang treatment [40] T^0_m (Eq. 9) is the equilibrium melting point for the pure polymer. In practice, the observed melting point for a crystalline polymer is hardly ever the equilibrium melting point. The dominant feature of crystalline polymers is the small size and, hence, large surface to volume ratio of the crystalline regions. These crystalline regions are so small that their surface energy is an important feature and this is primarily related to the thickness of the lamellae ($\sim$100 Å) which normally form the fibrils of the spherulites. The thickness of the lamellae depends on the crystallisation conditions and subsequent thermal history; thin lamellae have lower melting points but may thicken on annealing. Defects within fibrils can also reduce the melting points of crystalline regions.

On heating, thin or defective crystallites may melt and recrystallise at temperatures below the equilibrium melting point. Depending on the heating rates this initial melting may or may not be seen experimentally. Possibly just the final melting of the annealed material may be observed but even this process will not occur at the equilibrium melting point for thick lamellae. It has therefore become common practice to follow the procedure, described by Hoffman and Weeks (see Sect. 3.2) to determine values of T^0_m the equilibrium melting tem-

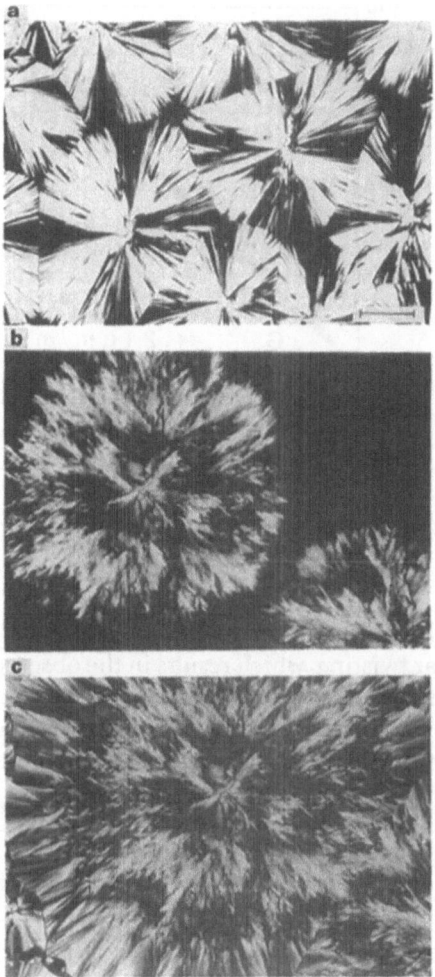

Fig. 8a–c. Optical micrographs of PCL spherulites in pure PCL crystallised at: **a** 45 °C for 30 min; **b** 55° C for 45 h; **c** 55° C for 45 h followed by crystallisation at 45° C; taken from [52]

perature at which the crystalline polymer is in equilibrium with the liquid phase.

The ring-banded structure of spherulites (Fig. 7) is often seen in blends in which the amorphous components are miscible, i.e. both components are present in the melt at the growth surface, and it is frequently considered that such features are only seen in these situations. Detailed information on ring-banded spherulites is exemplified in subsequent sections dealing with individual blend types. To date there is no agreed explanation for the origin of this feature; i.e. there is no explanation as to why the fibrils should twist in a regular

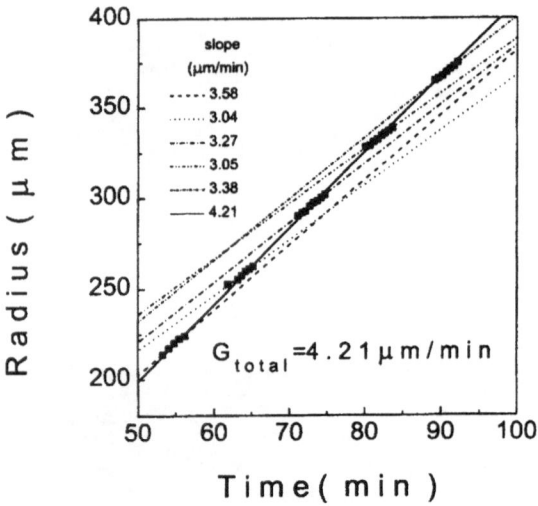

Fig. 9. Variation of the spherulite radius with crystallization time for PCL/SAN (90/10 w/w) blends with a free upper surface at 40° C Taken from [53]

manner at the growth surface. Recently, Kummerlöwe and Kammer [52] suggested that the lamellar twisting, which results in the observed periodicity in the spherulites, has its origin in the volume fraction ϕ of the amorphous (non-crystallising) component and the osmotic pressure at the growth surface. The osmotic pressure was given as $-\chi\phi^2$ and they suggested that the periodicity L could be written as

$$L\phi^k=(-\chi^{1/2})^m \tag{19}$$

where k (~1) and m are scaling parameters [52].

While it is often accepted that ring-banding in PCL spherulites only appears in blends, Kummerlöwe and Kammer published electron micrographs which question this view [52]. For pure PCL crystallised at 45 °C they reported well-formed spherulites with the classic Maltese Cross and no ring-banding, Fig. 8a. Electron micrographs of PCL crystallised at 55 °C showed far less regular structures with evidence of irregular but definite ring-banding, Fig. 8b. They also published an electron micrograph of a spherulite formed first at 55 °C and subsequently grown further at 45 °C (Fig. 8c). The centre was an irregular spherulite with evidence of ring-banding and there was an external ring of well-formed lamellar fibrils [52].

PCL spherulites, in pure PCL, have constant rates of radial growth in isothermal crystallisations. The same is often claimed for ring-banded spherulites in PCL blends. However, Wang et al. have recently cast doubts on this view [53]. These workers carefully monitored the growth of ring-banded spherulites and, while there is a constant average growth rate, found rhythmic growth rates (Fig. 9); growth during development of dark bands could not be monitored because of lack of light intensity in the microscope but the implication is that growth in such regions must have been faster than the average.

5
Natures of Blends

Blends may be formed by mechanically mixing (melt mixing) the different components or, because PCL is soluble in many solvents, blends may be produced from a solution of the components in a common solvent.

In melt mixing, for example in a Brabender Plasticorder or an extruder, shearing forces are applied to the mixture of polymers at temperatures above the T_g or T_m of the polymer with the highest softening or melting point. The objective is to disperse one component within the other, either to produce a fine dispersion of immiscible polymers or, if the components are miscible, to aid dissolution of one polymer into the other. For immiscible blends it might be anticipated that the texture of the dispersion would depend on the work expended on mixing, the size of the dispersed phase decreasing with increasing work done. It might also be expected that the minor (volume fraction) phase would be the dispersed phase but rheological factors can intervene and the relative viscosities of the components at the temperature of mixing can play a role. What is important is the combination of volume fractions and viscosities of the phases. Jordhamo et al. [54] have used, in PCL blends, the previously developed relationship

$$\alpha = \frac{\phi_B}{\phi_A} \frac{\eta_A}{\eta_B} \tag{20}$$

in which the ϕ and η terms are the volume fractions and viscosities of the respective phases. If $\alpha>1$ then phase B is predicted to be continuous and if $\alpha<1$ phase A continuous; if $\alpha=1$ then co-continuous phases are expected. This relationship is applicable if ϕ for the minor phase is >0.25 – otherwise the major phase is always the matrix phase.

Solvent mixing, less relevant commercially, is widely used in scientific studies to determine the natures of blends. By using dilute solutions of the components the polymers, miscible or immiscible in bulk, can be combined homogeneously. Slow removal of solvent from inherently immiscible polymer mixtures allows liquid-liquid phase separation to proceed and the polymers to segregate. However, rapid solvent removal or co-precipitation into a large volume of non-solvent can result in intimate mixtures of even immiscible polymers; results may depend on the solvent used. Thus, non-equilibrium, unstable mixtures of inherently immiscible polymers can be produced. Such mixtures may segregate when heated above the T_gs of the samples when molecular mobility permits. This situation is encountered many times in studies of PCL blends.

Solvent effects can influence the outcome of the casting process because of the intervention of polymer-solvent interactions in addition to the polymer-polymer interactions. Polymer-solvent interactions influence the viscosities of polymer solutions through their effects on the hydrodynamic volumes of polymer chains which are large in good solvents and small in poor solvents. Polymer-polymer interactions may compete with polymer-solvent interactions and modify the hydrodynamic volumes of polymer molecules and their connectivity through physical interactions (if favourable for miscibility), especially at high polymer concentrations, as solvent is removed. The only study relevant to this review appears to be a study on PCL/PVC blends and only in dilute solution; see Sect. 8.

Blending of polymers produces materials with some combination of properties which depend on the properties of the components, the extent of miscibility and the texture of phase-separated materials. Immiscible blends normally exhibit two T_gs, similar or identical to those of the separate components. In partially miscible systems those transition temperatures are modified, while in miscible blends a single, composition-dependent T_g is observed. In immiscible systems where the PCL, say, forms a separate phase it may crystallise. Under appropriate circumstances PCL may also crystallise from partially miscible or miscible systems. Thus, a variety of different types of blends can be achieved.

Often the T_gs of miscible (or partially miscible) blends are compared with values calculated from the Kelley-Bueche equation (Eq. 21) [55]. which is based on the additivity of free and occupied volumes of the individual components, the Gordon-Taylor equation (Eq. 22) [56] or the Fox equation (Eq. 23) [57]; for a binary mixture of A and B T_g is the glass-transition temperature of the mixture, ϕ_A, ϕ_B, w_A, w_B and $T_{g,A}$, $T_{g,B}$ are the volume and weight fractions and glass-transition temperatures of amorphous components A and B; in Eqs. (21) and (22) k is a constant given by Eq. (24) in which the α terms are coefficients of expansion and subscripts l and g refer to liquid and glass, respectively. Since, in practice, values of $(\alpha_l - \alpha_g)$ are similar for many materials, k often approximates to and is often assumed to be unity, when Eqs. (21) and (22) reduce to Eqs. (21a) and (22a). The increase in coefficient of expansion from glass to liquid is attributed to an increase in free volume which increases rapidly with temperature above T_g.

$$T_g = \frac{\left[\phi_A T_{gA} + k\phi_B T_{gB}\right]}{\left[\phi_A + k\phi_B\right]} \tag{21}$$

$$T_g = \phi_A T_{ga} + \phi_B T_{gB} \tag{21a}$$

$$T_g = \frac{\left[w_A T_{gA} + kw_B T_{gB}\right]}{\left[w_A + kw_B\right]} \tag{22}$$

$$T_g = w_A T_{ga} + w_B T_{gB} \tag{22a}$$

$$\frac{1}{T_g} = \frac{w_A}{T_{g,A}} + \frac{w_B}{T_{g,B}} \tag{23}$$

$$k = \alpha_A / \alpha_B = (\alpha_{1A} - \alpha_{gA}) / (\alpha_{1B} - \alpha_{gb}) \tag{24}$$

In order to predict values of T_g of mixtures it is necessary to have the correct values of T_g of the constituents. Because of the highly crystallisable nature of PCL it is difficult to obtain an unambiguous value of T_g for this polymer but it is possible to obtain an estimate of T_g by the use of modified forms of Eqs. (21) or (22); see below.

Many polymers which have been incorporated into polymer blends are amorphous. However, several, including PCL, crystallise, usually in the form of spherulites. We have already remarked that such polymers are never 100% crystalline; amorphous and crystalline components coexist and, on heating, exhibit both a glass-transition temperature, for the amorphous component, and a melting point (or melting points), for the crystalline component (or components). Values of T_m may vary with the quality of the fibrillar crystalline regions and equilibrium values may be difficult to determine. Also, amorphous chains (such as tie chains) within the spherulites may have restricted molecular motions and not exhibit the normal T_g for that polymer.

Using blends containing PCL, Koleske and Lundberg extended previous studies of polymer blends of wholly amorphous polymers to ones in which one component is crystalline [2]. An inherently crystallisable component may crystallise from a miscible amorphous blend, in a liquid-solid phase separation, and exhibit a melting transition close to that of the homopolymer; normally the crystalline material would be a pure single polymer; it is feasible that both components of a binary blend may crystallise separately.

For polymer blends in which one component is crystalline the melting behaviour depends on circumstances. For immiscible blends, where the components are phase separated (prior to crystallisation) and act independently, the crystal melting temperature will be that of the homopolymer. In miscible blends, where the amorphous phase contains both components, the melting temperature will be lower than the equilibrium melting temperature for the crystallisable homopolymer, i.e. the crystalline polymer exhibits a melting point depression as discussed above. The Nishi and Wang approach (Sect. 3.2) has been used to estimate the magnitude of the interaction parameters in a number of blends (Sect. 7). Poly(ε-caprolactone) blends are often semi-crystalline and the above considerations, therefore, apply to many PCL blends.

In semi-crystalline blends it should be noted that when the value of T_g of the amorphous phase is compared with values in Eqs. (21)–(23) the observed T_g reflects the composition of the residual amorphous phase rather than that of the overall mixture. Currently, the most common method of determining the crystalline content of a blend is to compare the energy associated with the melting endotherm, in, for example, a DSC thermogram, with that for the homopolymer and its previously established crystalline content.

The value of T_g of the amorphous component in a pure partially crystalline polymer may also differ from that of the wholly amorphous material because of restrictions on chains closely associated with the crystalline regions, in loops or tie-chains, etc.; this is true for PCL and different authors have reported different values of T_g for samples prepared under different conditions.

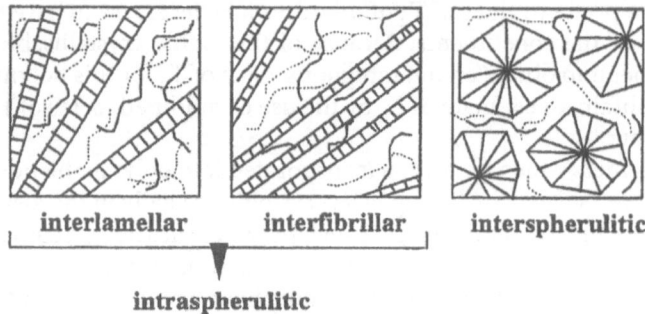

Fig. 10. Schematic representation of segregation of amorphous components in partially crystalline polymer blends depicting the location of residual crystallisable polymer and non-crystallisable polymer in the interlamellar and interfibrillar regions within spherulites and interspherulitic locations. *Solid lines* represent the crystallisable component and *dotted lines* the non-crystallisable component; taken from [60]

In order to acquire a value of T_g for use in Eqs. (21)–(23), it is common practice to quench samples from homogeneous mixtures at high temperatures in order to hinder crystallisation and to produce wholly amorphous polymer at temperatures below the T_gs of the mixtures. In this way Koleske and Lundberg obtained a value of 213 K (−60 °C) from dynamic mechanical studies of quenched PCL homopolymer but a value of 202 K (−71 °C) by extrapolation of dynamic mechanical properties of blends with PVC, using a linearised form of Eq. (22) from which it is possible to obtain estimates of k and T_{gA}. For that system they obtained a value of 0.519 for k [2]. The linearised form of Eq. (22) is

$$T_g = T_{gA} + k\left[\left(T_{gB} - T_g\right)w_B / w_A\right] \tag{25}$$

where polymer A is the crystallisable component (e.g. PCL) for which an unambiguous value of T_g cannot be readily obtained. Similarly, Hubbell and Cooper obtained values for PCL T_g and k of 187 K (−86 °C) and 0.786 and 182 K (−91 °C) and 1.57 from quenched samples of PCL blends with poly(vinyl chloride) and nitrocellulose, respectively; quenched pure PCL gave a T_g of about 210 K (−63 °C) [58].

We have already remarked (Sect. 4) that even in a crystallisable homopolymer there is a non-crystalline component which can be located in different situations. Similarly, in a blend of a crystallisable polymer and a non-crystallisable polymer, the non-crystallising component may be located in different situations. That polymer may be retained within the amorphous component of the spherulites, either between lamellae or between fibrils. Or the polymer may be rejected from the spherulites and located in an amorphous phase between the spherulites; in this case the spherulites will not be space-filling. These several possibilities have been discussed for blends of poly(ε-caprolactone) by Defieuw et al. [59]. They have also been depicted schematically by Vanneste et al. (Fig. 10) [60].

It is also possible that, in blends, the amorphous component of the spherulites might be homogeneous or phase separated. Non-crystalline material of the crystallisable component at the surface of the lamellae might be sufficiently ordered that the other (non-crystallising component) might be concentrated into a more restricted volume or might even phase separate.

6
Polymeric Plasticizers

Poly(ε-caprolactone), having a very low glass-transition temperature, is usually blended with polymers of higher glass-transition temperature and, in miscible systems, it therefore plasticizes (i.e. reduces the T_g of) the other component. Plasticizers are technologically important in imparting flexibility to hard plastics, e.g. to render PVC flexible, but low molecular weight plasticizers, which are soluble materials with low T_g and high free volume content, often suffer the disadvantage that they may be volatile and lost slowly from the matrix polymer; Eqs. (21)–(23) also apply to such systems. PCL, therefore, is potentially valuable for such purposes in that it can act as a useful polymeric plasticizer (it has a high associated free volume) which is non-volatile and, hence, is not susceptible to loss by slow evaporation [61].

Mullins, in 1966, patented the use of PCL and other solid rubbery lactone polymers as polymeric plasticizers for vinyl polymers, especially for poly(vinyl chloride) [1]. He claimed outstanding low-temperature impact strength for blends with 25–125 parts of PCL per 100 parts of PVC resin; the products had flexibility below 0 °C and embrittled at temperatures as low as –65 °C. The plasticizers claimed were high molecular weight, solid, tough, rubbery polymers from cyclic esters (polymers and copolymers), with reduced viscosities preferably between 1.2 and 4, determined as 2% solutions in chloroform. The blends were reported to be tougher, and more extensible than those prepared with conventional plasticizers. The original patent notes the low volatility of PCL plasticizer (loss of volatiles was reported to be negligible at 70 °C) and reports higher resistance to extraction by oil and water than encountered with commercial polyester plasticizers. In addition, it was reported that the PCL blends have better drape, softness and hand.

Plasticization is often achieved by mechanical mixing of preformed components but this is not the only way of combining the components. In addition to simply mixing PCL (the low T_g component) with PVC to plasticize the PVC, a claim has been made for the preparation of internally plasticized PVC by polymerising vinyl chloride in the presence of preformed PCL [62]; the claim is said to produce transparent, plasticized PVC suitable for use as a packaging film.

Several claims have been made for the use of PCL as a plasticizer for other polymers which can then be used as plastic clays for modelling purposes. Thus, 5–50 parts of PCL are said to plasticize chlorinated polyethylene (25–45 wt % Cl) to produce plastic clays with softening points of 40–60 °C [63, 64]. Similarly, blending PCL with styrene-acrylonitrile copolymer is said to produce materials which soften above 50–60 °C [65] and to produce mouldable materials when heated to 40–60 °C by blending with bisphenol-A polycarbonate [66]. In addition, ductility of polycarbonate/PVC blends is said to be enhanced by the addition of PCL [67]; PCL is said to be miscible with each of these components. The general behaviour of blends of PCL with each of the cited polymers is discussed in subsequent sections.

7
Interactions and Interaction Parameters in Poly(ε-caprolactone) Blends

Interactions between species are often reflected in spectroscopic features and Garton [68] has summarised relevant features which are apparent in infrared spectroscopy of a number of polymer blends, composites and surfaces. In that survey Garton discussed both practical and theoretical aspects of interpreting infrared spectra of polymer blends. Naturally, homopolymers have intergroup interactions which may be replaced by any interactions which might develop between unlike species in blends. There is little self-association of carbonyl groups in pure polyesters. Consequently, any development of H-bonding in PCL blends which involves the PCL carbonyl groups lowers the carbonyl stretching frequency. Hydroxyls in poly(vinyl phenol), for example, can self-interact so their association with carbonyl can give an increase or decrease in the hydroxyl stretching frequency, depending on relative strengths of the several interactions which might occur.

Coleman and Zarian [15] investigated PCL/PVC blends (PCL-700; PVC QYT-387, Table 1). Samples were cast from THF; those containing <75 wt % PVC were dried in air and those with >75% PVC were dried in a vacuum desiccator at 60 °C. The samples were examined by FTIR and direct evidence of interactions involving carbonyl groups was obtained; the C=O stretch was found to shift to lower frequency as a function of PVC concentration and widths at half height of absorptions increased in the melt (70 °C). Effects saturated at about 4:1 PVC/PCL mole/mole (60 wt % PVC), which is consistent with the relative sizes of repeat units in the polymers and development of maximum concentration of interactions; the PCL repeat unit is about 3.4 times longer than the PVC unit. In semicrystalline blends at high PVC content (>2:1 PVC/PCL mole/mole), where the blends are amorphous, very similar FTIR data in terms of peak shifts and widths were observed in the solid and molten states, indicating the existence of similar interactions in both situations. At mole ratios up to 2:1 spectra were consistent with the coexistence of crystalline and amorphous PCL phases. Also apparent was a shift in C=O stretching frequency in the amorphous component of the combined peak. It has been pointed out that such small shifts might be ambiguous; apparent shifts might result from overlap of shifted and unshifted peaks so that quantitative determination of fractions of carbonyl involved in intermolecular interactions might be incorrect.

It is generally considered that the interaction between PVC and PCL, or other polymers containing carbonyl or ether linkages, arise through weak hydrogen-bonding between the α-H of PVC and carbonyl.

$$\begin{array}{c} \qquad\qquad\quad | \\ \diagup C = O - - - H - C - Cl \\ \qquad\qquad\quad | \end{array}$$

However, measurements of heats of mixing between PVC or chlorinated hydrocarbons and THF or other ethers by Pouchly and Biros [69] showed that heats of mixing depended primarily on Cl content rather than molecular structure, i.e. there was no correlation with H-C-Cl content. Nevertheless, it was stated by Walsh and Rostami [34] that Varnell [70] obtained evidence, from studies with deuterated PVC, that the α-hydrogen in PVC is involved in the intermolecular interactions. Further, Prud'homme [71] suggested that dipole-dipole interactions between carbonyl and C-Cl groups, found in mixtures of liquid esters and PVC by Pouchly and Biros [69], are important.

Inverse-phase gas chromatography (Sect. 3.4) is now used frequently to assess interaction parameters and was used by Riedl and Prud'homme to estimate polymer-polymer interaction parameters in polyester-PVC blends, including PCL-PVC blends, at 120 °C and as a function of polymer concentration [72]. Interaction parameters were found to vary with the CH_2:ester ratio in the polyester, i.e. with n in 2. From retention volumes they calculated values of the interaction parameter χ_{AB} (the Flory interaction parameter per segment) as functions of blend composition with various (usually six) probes. The exercise was performed for several polyesters with different CH_2:ester ratios. Values of the interaction parameters varied from probe to probe but showed similar patterns (Fig. 11). Mean values of the parameters were calculated for each system at each composition (Table 2). (The original paper contains a considerable amount of tabulated data.) For the various polyesters χ_{AB} was uniquely negative (favourable for miscibility) for CH_2:CO_2=5, i.e. for PCL, at all volume fractions. The mean values of the interaction parameters varied with PCL content with large negative values at high PCL contents and values near zero at low PCL contents, consistent with a stronger driving force for miscibility at high PCL content; the interaction parameter was insensitive to composition from 75 wt % to 0% PCL where χ_{AB} was only just negative (near zero). The variation χ_{AB} with CH_2:ester ratio is shown for volume fractions of PCL of 0.5 and 0.8 (Fig. 12); a one to one C=O to Cl ratio occurs at a PCL volume fraction of 0.7 and the results are tabulated in [72].

In an earlier study of PCL-PVC blends by the same technique, Olabisi [18] also found negative interaction parameters for PCL-PVC mixtures using several probes and concluded that LCST behaviour was to be expected in the phase diagram with a minimum in the miscibility curve at relatively high PVC concentrations, in agreement with data reported above and elsewhere (see also Sect. 8 on blends with PVC).

The results obtained by Riedl and Prud'homme indicate that of all the aliphatic polyesters, PCL is unique in being miscible with PVC although the same types

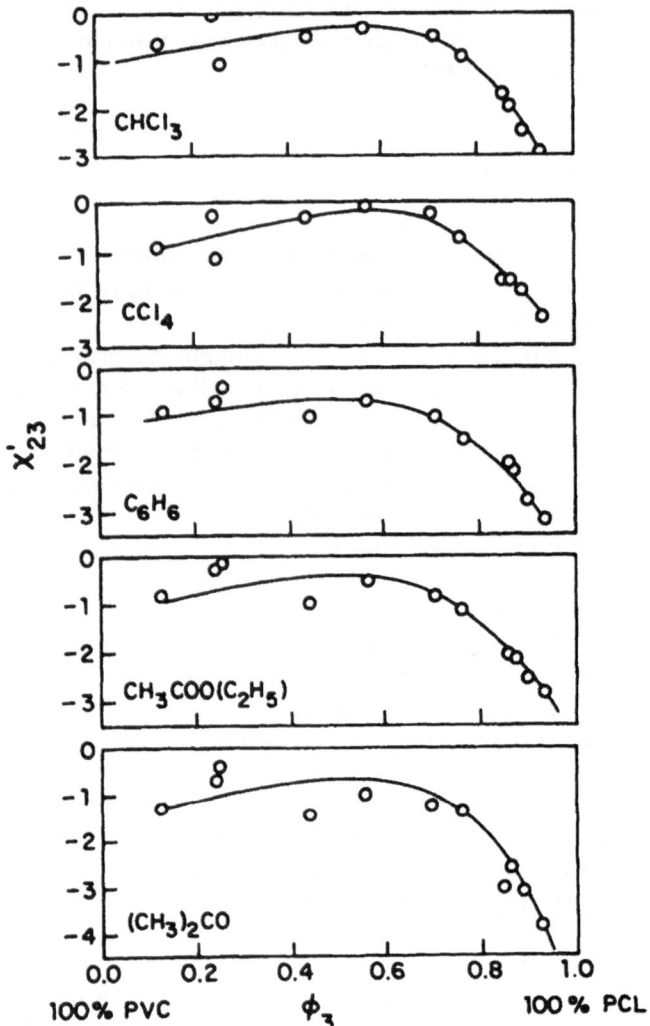

Fig. 11. Values of the polymer-polymer interaction parameters (χ), determined by inverse-phase gas chromatography with different probes, for PVC/PCL blends as functions of PCL volume fractions; taken from [72]

of interactions are available in all such polyesters [72]. Miscibility with PCL blends must therefore be due to a favourable balance of interactions in which the more common unfavourable interactions are outbalanced by the content and strength of the favourable interactions. Woo et al. [73], attributed immiscibility at high carbonyl contents, i.e. low values of n in 2, to an unfavourable balance of inter- and intramolecular interactions, between methylene and carbonyl. Prud'homme [71] stated that immiscibility at low carbonyl contents, i.e. high val-

Table 2. Mean values of polymer-polymer interaction parameters (χ), determined for several probes, in PCL/PVC blends determined from inverse-phase gas chromatography at 120° C; data taken from [72]

Volume fraction of PVC	Interaction parameter
0.064	−2.6
0.100	−2.3
0.127	−1.7
0.144	−1.7
0.231	−0.8
0.291	−0.5
0.435	−0.2
0.552	−0.5
0.740	−0.2
0.755	−0.1
0.868	−0.6

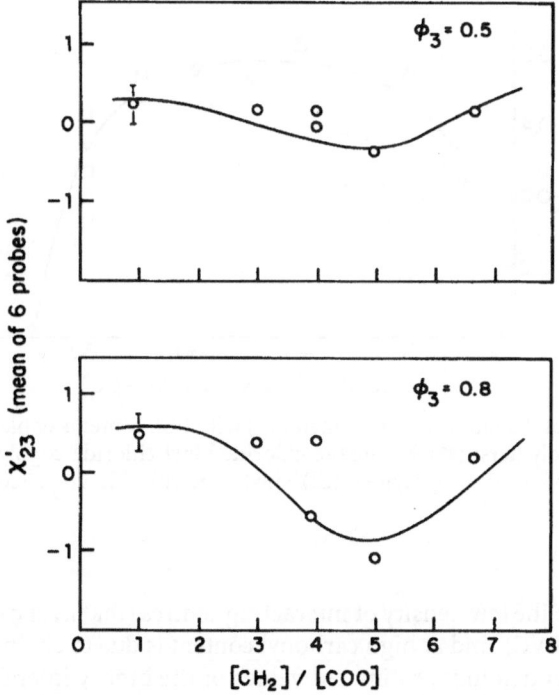

Fig. 12. Variation of the polymer-polymer interaction parameter (χ) for blends of PVC with several polyesters as functions of the polyester CH_2/COO ratio; taken from [72]

Table 3. Interaction parameters (χ) for blends of PCL with vinylidene chloride copolymers [74]

PCL content (wt %)	Vinylidene chloride comonomer		
	vinyl chloride	vinyl acetate	acrylonitrile
95			−3.41
90	−1.06	−1.94	
80	−0.46	−0.53	−0.37
70	−0.32	−0.26	−0.28
60	−0.25	−0.21	−0.29
50	−0.18	−0.18	−0.28
40	−0.16	−0.19	
30	−0.18	−0.18	
20	−0.25	−0.26	
10	−0.55	−0.47	

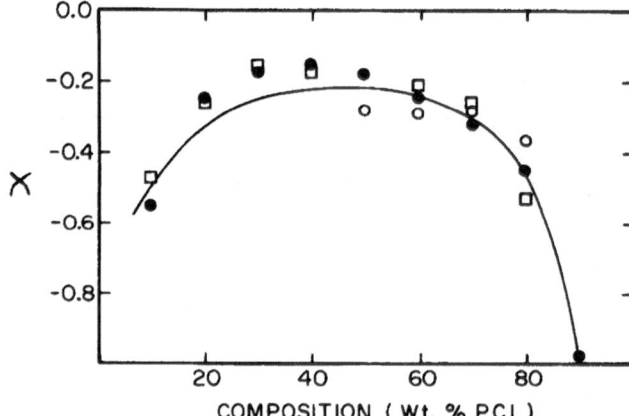

Fig. 13. Variation of the interaction parameter (χ) with PCL contents of blends of PCL with several Saran copolymers: (●) vinylidene chloride-vinyl chloride copolymer, (○)vinylidene chloride-acrylonitrile copolymer, (□) vinylidene chloride-vinyl acetate copolymer; data taken from [74]

ues of n, is due to the low density of interacting centres (that is, of carbonyl groups to interact with PVC), and at high carbonyl content is due to chain stiffness.

Zhang and Prud'homme estimated values of the binary interaction parameters for PCL with vinylidene chloride copolymers known as Saran, that is with copolymers of vinylidene chloride with, separately, vinyl chloride, acrylonitrile or vinyl acetate [74]. Interaction parameters were estimated from melting point depressions. Values obtained were found to be composition-dependent and are tabulated in Table 3; the composition dependence is also apparent from the data in Fig. 13.

From studies of styrene-acrylonitrile copolymer (SAN) and poly(methyl methacrylate) (PMMA) blends, Higashida et al. [75] used previously determined values of the interaction parameter between styrene (S) and acrylonitrile (AN) units [76] and phase diagrams of PCL/SAN blends obtained by Chiu and Smith [20] to estimate values of $\chi_{S/PCL}$ and $\chi_{AN/PCL}$ as functions of temperature and AN content. On this basis they calculated phase diagrams similar to that determined experimentally by Chiu and Smith [20]. They also predicted that mixtures of PS and PCL oligomers should be miscible at low temperatures and that such mixtures should exhibit UCST behaviour which was in agreement with experimental observations; values of the interaction parameters determined by Higashida are quoted in Table 4.

Subsequently, Svoboda et al. [77] undertook FTIR and calorimetric measurements on low-molecular-weight analogues to determine the existence and magnitude of the specific interaction between units in blends of PCL and SAN. As analogues, they used ethylbenzene (EB) for styrene, propionitrile (PN) for acrylonitrile and n-propyl propionate (PP) for caprolactone and hexane (H) as a noninteracting diluent. They observed decreases in the C=O stretching frequency for PP in presence of EB, and more so with PN. There was also an increase in the out-of-plane C-H bending frequency for the benzene ring in EB in the presence of PP or PN and a decrease in the C≡N stretching frequency of PN in presence of EB or PP. These shifts indicate the existence of several interactions between the monomeric units in PCL/SAN blends.

Microcalorimetric measurements were undertaken on various mixtures to determine heats of mixing. Endothermic heats of mixing for the pairs PN+EB and PN+PP were measured, consistent with the known immiscibility of PAN with PCL and PS with PAN. Exothermic mixing was found for EB+PP and this result disagrees with the observed immiscibility of PCL with high-molecular-weight polystyrene. Svoboda et al. described the procedure used to analyse the data and to determine interaction parameters. They list the self-interaction and association energies for relevant species and pairs of units; data are tabulated in the original reference [77]. Good agreement was found between the enthalpies of mixing measured, using low-molecular-weight analogues, and those calculated with variations in the mole fraction of AN in the SAN, showing the self-consistency of the model used for data analysis. These workers also compared their heat of mixing data with the established phase diagram for PCL/SAN blends,

Table 4. Interaction parameters ($\chi=A+B/T$) for blends of PCL with polyacrylonitrile and polystyrene; data taken from [75]

Polymer	A	$10^2 B$
polyacrylonitrile	−4.76	21.4
polystyrene	−0.913	4.12

which showed a miscibility window for blends with copolymers having AN contents of about 15–45 wt %, with immiscibility at high (>85%) styrene contents. The heat of mixing data were consistent with immiscibility at high AN, where a positive ΔH_m was observed, and ΔH_m became negative at about same composition as the onset of miscibility. However, the calculated values of ΔH_m continued negative to blends with pure polystyrene, which disagrees with known immiscibility of PCL and high-molecular-weight polystyrene [77]. They also undertook some measurements using ethyl butyrate as an alternative PCL analogue and again observed negative heats of mixing with EB, confirming negative enthalpy of mixing and the existence of favourable interactions with polystyrene. These results are in conflict with the positive value for $\chi_{PCL/PS}$ observed by others [78]. The origin of this discrepancy is not clear. Svoboda et al. list four factors involved in determining values of χ-parameters in equation of state theories which might be responsible: they are exchange energy, free volume, size effect and other non-combinatorial effects. It was considered that other effects might be able to overcome the negative exchange energy term and result in overall unfavourable interactions in mixtures of PCL with polymers rich in styrene. Another possibility is that end-group effects might not be negligible, so that the analogues are not truly representative of polymer units [77]. It is notable that PCL has been found to be miscible with polystyrene oligomers; see Sect. 17.1.

Interaction parameters between PCL and SAN of different compositions, i.e. with different AN contents, have been estimated from crystallisation studies and these are tabulated in Table 5 [79]. In addition, Shah et al. determined a value of B for blends of PCL with SAN-25 of -2.55 J cm^{-3} [43]. Li et al. quoted a value for χ in blends with SAN containing 24 wt % AN of -0.33 [80].

Li et al., from a knowledge of the phase diagram for PCL/polystyrene blends, determined a value of the Flory-Huggins interaction parameter χ for this polymer pair of $0.061+150(RT)^{-1}$ J mol^{-1} [81], while Watanabe et al. quoted a value of $\chi=(-469/32)(1-2.46\times10^4/T)$ [82].

There have been several studies of interactions between PCL and phenoxy resin, a polymer prepared from bisphenol-A and epichlorhydrin, and several estimates of the interaction parameter have been made; see Sect. 16. Harris et al. estimated a value for B (Eq. 8) of -10.1 J cm^{-3} from group contributions and attributed this to a favourable interaction between the -OH of phenoxy and the carbonyl groups of PCL, counteracted by other interactions [83]. De Juana et al. [84] used inverse gas chromatography and found values of B between -4.8 and -16.1 J cm^{-3}, depending on composition, temperature and method of estimation. Coleman and Moskala [85], in FTIR studies, identified a peak for carbonyl

Table 5. Interaction parameters between PCL and SAN copolymers with different AN contents (n wt %) in SAN-n copolymers [79]

Wt % AN	12.4	14.9	19.5	21.9	26.4
$10^3\chi_{AB}$	-3.727	-4.583	-5.199	-5.116	-4.335

hydrogen-bonded to the hydroxyl of phenoxy. These workers concluded that this hydrogen-bonding interaction was weaker than that in phenoxy itself (see Sect. 16).

For blends of PCL with bisphenol-A polycarbonate, Shah et al. determined a value of the interaction parameter B of -1.6 J cm^{-3} [43]. Values of interaction parameters B with tetramethylbisphenol-A polycarbonate of -7.5 J cm^{-3} [86] and -8.2 J cm^{-3} [87] have been reported.

FTIR studies have shown the development of an additional carbonyl stretching vibration for PCL carbonyl at 1725 cm^{-1} in blends of PCL with phenol formaldehyde resin [88].

FTIR studies of PCL blends with poly(4-hydroxystyrene) showed the presence of a band at 1708 cm^{-1} which varied in intensity with the content of poly(4-hydroxystyrene) and was attributed to favourable interactions between the carbonyl groups of PCL and the hydroxyl groups of poly(4-hydroxystyrene); see Sect. 17.3 [89]. From crystallisation studies, Lezcano et al. [42] determined a value for χ_{AB}/V (where V is a molar volume for the repeating units and taken as 100 cm^3 mol^{-1} for 4-hydroxystyrene) of -1.3×10^{-2} mol cm^{-3} (without allowing for entropic contribution and -1.1×10^{-2} mol cm^{-3} allowing for this contribution) which agreed with values between -7×10^{-3} mol cm^{-3} (ϕ_{P4HS}=0.23) and -1×10^{-3} mol cm^{-3} (ϕ_{P4HS}=0.72) obtained by inverse gas chromatography and which extrapolated to -1.2×10^{-2} mol cm^{-3} at ϕ_{P4HS}=0 [90].

For blends of PCL with poly(4-hydroxystyrene) in which 60% of the hydroxyl groups were methoxylated to provide a copolymer of 4-hydroxystyrene and 4-methoxystyrene (Sect. 17.4), a value for χ_{AB}/V of -0.83×10^{-2} mol cm^{-3} was obtained which is lower than the value for blends with pure poly(4-hydroxystyrene) and consistent with the reduced content of hydroxyl groups which favour polymer-polymer interactions through hydrogen-bonding [91]. The absorption found at 1708 cm^{-1} in blends of PCL and poly(4-hydroxystyrene), assigned to the hydrogen-bonded carbonyl group, was also present in the blends with the 4-hydroxystyrene and 4-methoxystyrene copolymer [91].

Barnum et al. determined values of B for blends of PCL with copolymers of styrene and allyl alcohol with different hydroxyl (allyl alcohol) contents of 1.3–7.7 wt % [92]. They found values of B to vary with hydroxyl content but to be between about -4 J cm^{-3} and -12 J cm^{-3}; see Sect. 17.2.

From phase diagram data, Watanabe et al. determined for blends of PCL with poly(vinyl methyl ether) a value for the interaction parameter of $\chi=(-9.32/1.8)(1-396.5/T)$ [82].

8
Blends with Poly(vinyl chloride) (PVC)

Blends of PCL with PVC 3 have been studied by several groups. The first report by Mullins [1] described the use of PCL as a polymeric plasticizer, as discussed in Sect. 6. In the initial studies of the nature of PCL blends, Koleske and Lundberg [2] established the essential features of the blends which have been subsequently confirmed by many others. Samples were prepared by melt mixing PCL with a viscosity-average molecular weight of 41,000 and a commercial PVC sample from Union Carbide (QYTQ, Table 1). They demonstrated that blends rich in PVC exhibited a single, composition-dependent, glass-transition temperature indicating miscible blends. Blends containing 50 wt % or more PCL were shown to be partially crystalline. That is, at high concentrations of PCL part of the PCL is phase-separated by crystallisation.

$$\left[\text{CH}_2 - \overset{\overset{\displaystyle \text{H}}{|}}{\underset{\underset{\displaystyle \text{Cl}}{|}}{}} \right]_n$$

Structure 3

Using their extrapolated value of T_g for PCL of –71 °C, Koleske and Lundberg found that the T_gs of the blends varied with composition and the data gave good fits with both the Fox and Gordon-Taylor equations (Eqs. 23 and 22, respectively) [2]. Thus, blends with less than about 50 wt % PCL were homogeneous and exhibited a single T_g; the blends were soft and pliable because the inherent PCL crystallinity was destroyed and the PVC was plasticized by the amorphous PCL. At 30% or less PCL amorphous blends were stable for long times [2]. Blends with more than 40% PVC, which were initially homogeneous, became heterogeneous as PCL crystallised. Under such conditions the tensile modulus was time dependent. The modulus increased slowly at 40 wt % PCL over a period of one month, increased much more rapidly at 45 wt % and 50 wt % PCL, over two to five days, and continued to increase slowly over long periods of time. Thus, the elastic modulus decreased with increasing PCL content in amorphous systems where PCL acted as a plasticizer and increased (at long times) at higher PCL content with increasing content of crystalline PCL which acted as a reinforcing agent (Fig. 14).

In blends the storage shear modulus showed a major decrease at T_g of the blends, at decreasing temperatures with increasing PCL, but in blends containing more than 50 wt % PCL, after falling at T_g, the modulus rose again as the temperature increased and PCL segregated by crystallisation [1].

Hubbell and Cooper [58] obtained similar data on T_gs of PCL/PVC blends using DSC measurements. They used their value of about –90 °C for T_g of wholly amorphous PCL, approximated to using quenched samples, and showed that T_gs for the blends were again consistent with both the Fox (Eq. 23) and Gordon-Taylor (Eq. 22) equations; quantitatively the data of Koleske and Lundberg [2] gave a better fit. Hubbell and Cooper also noted that the T_gs of annealed, partially-crystalline samples containing more than 50 wt % PCL were higher than those of the quenched amorphous samples [58], consistent with a portion of the PCL being crystalline and the PVC content of the residual amorphous phase being greater than that of the overall PVC content, thus raising the T_g of the amorphous component.

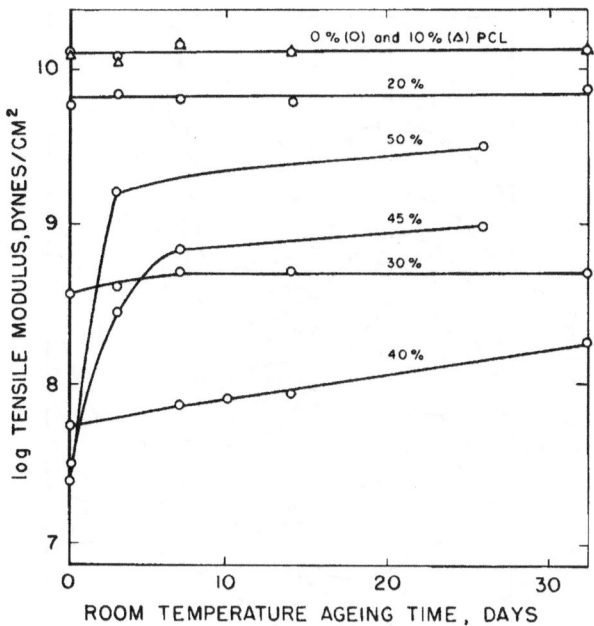

Fig. 14. Variations in tensile modulus with ageing time for PCL blends with PVC with different PCL contents (wt %); data taken from [2]

Modifications to secondary relaxations in polymers can be of practical importance because these relaxations are often correlated with toughness. Koleske and Lundberg [2] studied secondary relaxations in PCL/PVC blends, using loss modulus data (Fig. 15). Each homopolymer showed a β-relaxation with loss peaks (T_β) for PCL at –135 °C and for PVC at –40 °C; in PCL the β-relaxation was attributed to motions in the sequence of methylene units. Adding PCL to PVC caused the PVC relaxation to shift to a lower temperature as a result of effective plasticization of molecular motions in the PVC by the low-T_g PCL. Application of the Fox equation (Eq. 23) to secondary loss peaks in the blends, in a similar manner to its use in α-transitions (glass transitions), gave a good description of variations in properties for polymers containing more than 50 wt % PVC. The β-relaxation in PCL however did not appear to shift noticeably in blends rich in PCL but for blends containing more than 50 wt % PVC the relaxation shifted to lower temperatures (to about –148 °C at 90% PVC); the origin of this shift was not explained.

Barron et al. [23] also studied loss peaks in PCL/PVC blends containing more than 70 wt % PCL (Tone P787, Table 1), in this case by dielectric relaxation at 10 kHz; PVC molecular weight was about 110,000. They observed a β-relaxation peak at about –50 °C and a γ-relaxation peak at –85 °C in pure PCL. For blends with very high PCL contents (90–98 wt %) they found the temperature for the maximum of the relaxation peaks to be composition invariant, the β-relaxation

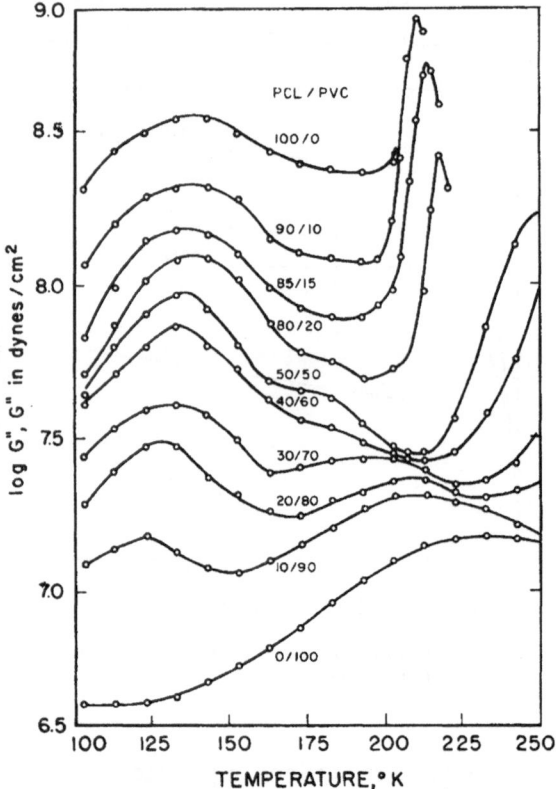

Fig. 15. Temperature dependencies of shear loss modulus for PCL-PVC blends, showing the secondary relaxations, curves shifted for clarity; data taken from [2]

peak for pure PCL decreasing in intensity on addition of PVC. In addition, for blends containing 98 wt % or less PCL a new composition-dependent peak grew to become a major loss peak at 25 °C for 70 wt % PCL, from 0 °C at 98 wt % PCL. They suggested that the composition-invariant β-relaxation was due to pure in-terphase PCL in the blend and attributed the new peak to the formation of an ad-ditional mixed phase; i.e. they assumed that there are two amorphous phases in blends containing about 90 wt % PCL. The activation energies for the β-relaxa-tion of the mixed phase and the γ-relaxation were found to be ~80–120 kJ mol^{-1} and ~40 kJ mol^{-1}, respectively, which are low compared with the interphase re-laxation activation energy of 100 kJ mol^{-1}. They claimed that dielectric experi-ments gave better resolution at 10 kHz and that, because of the frequency de-pendence of the peaks, the mixed and interphase relaxations overlapped at lower frequencies, characteristic of dynamic mechanical or DSC studies (i.e. the pres-ence of the interphase material was obscured) and were therefore not observed by other workers.

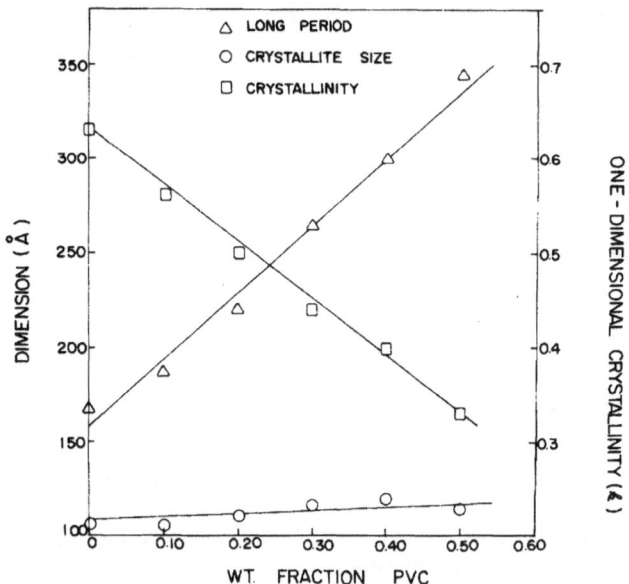

Fig. 16. Variations in long spacing, crystallite size and crystallinity (k) with PVC content in PCL-PVC blends; taken from [16]

Ong and Price [93] demonstrated that blends, cast from MEK (dried at room temperature under vacuum for 72 h) and high in PCL (Union Carbide, $\overline{M}_w$= 86,000) (65 wt % or more) contained PCL spherulites which filled space. Spherulites (which had a banded structure) in blends were found to be slightly larger than in pure PCL and their number decreased as the PCL content decreased; that is, a smaller number of spherulites were nucleated at lower PCL contents. Ong and Price found that at 50 wt % PCL the spherulites were not quite space filling. Khambatta et al. [16] also observed an increase in long-spacing (lamellar thickness plus separation of lamellae) as the PVC content increased (Fig. 16). The latter group confirmed that the spherulites were not space-filling at lower PCL contents. Khambatta et al. suggested that at low PCL contents the PCL is molecularly distributed within the PVC [16], and that the failure of PCL to crystallise was possibly because the PCL regions were too small to nucleate crystal formation; also there would be an effective reduction in melting point and reduced rate of crystallisation. Scattering studies (SAXS) suggested that the amorphous interlamellar regions were not homogeneous but phase separated into different phases of mixed composition, possibly separated by a transition zone of the order of about 30 Å [16].

Extensive crystallisation studies of PCL/PVC blends were undertaken by Ong and Price [94] (PVC $\overline{M}_w$=86,000, $\overline{M}_w/\overline{M}_n$~2); cast from MEK (dried at room temperature under vacuum for 72 h)) and by Stein and coworkers [9, 16] (PCL-700; PVC QYTQ-387, Table 1). The latter group used samples (solvent cast from

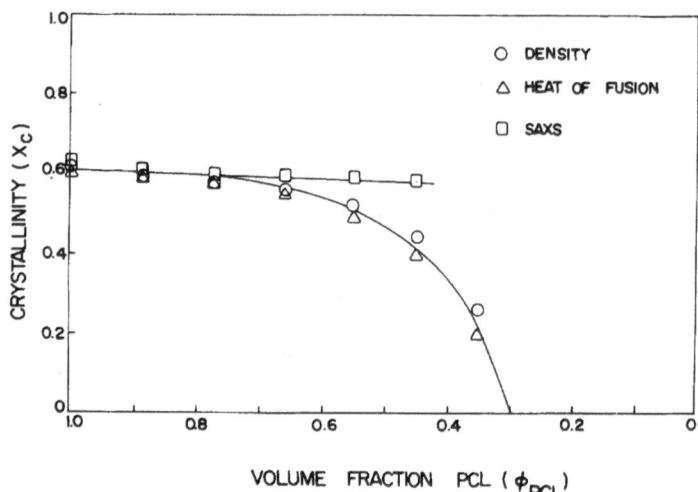

Fig. 17. Variations in crystallinity in PCL-PVC blends, as functions of PCL content, determined by different techniques; taken from [16]

THF and isothermally crystallised by heating to 70 °C and storing at 30 °C) which had a lower crystallinity (60%) than the samples prepared by Ong and Price (69±4%) by solvent-casting from MEK. Ong and Price restricted their study to blends containing more than 50 wt % PCL because below that content (as explained above) rates of crystallisation were very low. Khambatta et al. [16] determined degrees of crystallinity using the same specific volumes as Ong and Price of 1.390 g cm^{-3} for PVC and 1.094 g cm^{-3} and 1.187 g cm^{-3} for amorphous and crystalline PCL, respectively. They found about 60% PCL crystallinity with 75 wt % PCL in the blend, which fell to about 40% at 45 wt % PCL and to zero at 30 wt % PCL, by heat of fusion and density measurements. However, PCL crystallinity was said to be retained at 45–60% of the PCL when determined by SAXS measurements (Fig. 17). Ong and Price [93] stated that PCL crystallinity decreased with increasing PVC content; their measurement of 69% crystallinity in pure PCL was higher than the value 48% quoted by Crescenzi et al. [13] for samples cast from THF. The latter group found that crystallinity decreased to 50% at 50 wt % PCL, then fell severely to 30% at 40 wt % PCL and to zero at 30 wt % PCL. Possibly the casting solvent had some influence on crystallinity which could explain the quantitative differences between the observations of the two groups (see discussion of influence of experimental conditions on the properties of blends with SAN, Sect. 11).

Nojima et al. [95] studied the crystallisation of PCL ($\overline{M}_w$=23,300, $\overline{M}_w/\overline{M}_n$= 1.58) alone and in blends with PVC ($\overline{M}_w$=144,000, $\overline{M}_w/\overline{M}_n$=1.56) containing 90 wt % PCL by small angle X-ray scattering (SAXS). Samples, initially cast from THF, were heated to ~65 °C, to melt the PCL, and then quenched to the crystallisation temperature (28–43 °C); cooling was achieved in 60–100 s and crystallisation started during cooling. For pure PCL the peak of scattering intensity for samples crystallised at T_c=40.9 °C corresponded to a repeat distance (lamellae plus amorphous layer) of 16.6 nm, approximately independent of T_c. The long-spacing in the blend was slightly less than that in pure PCL and increased slowly with T_c to equal that for PCL at T_c~40 °C. Thus the internal texture of the spherulite in the blend was slightly different to that in pure PCL. Although the repeat distance in the blend was slightly smaller no evidence was presented to indicate the relative changes in the crystalline lamellar and amorphous layer thicknesses individually. This difference was coupled with the non-banded and banded structures seen for the pure PCL and blend, respectively. Small changes in long-spacing in the early stages of crystallisation were attributed to reorganisation of lamellae or formation of additional lamellae in inter-lamellar spaces.

The intensity of the scattering peak increased with time in a sigmoidal manner to plateau at a value I_∞, corresponding to a limiting PCL crystallisation after about 1200 s for both pure PCL and for the blend at temperatures of about 40 °C. From values of I_∞ times $\tau_{1/2}$ required to reach $I_\infty/2$ were calculated. Values of $\tau_{1/2}$ showed that PCL crystallisation in the blend was slower than in pure PCL and that the rates of crystallisation, as might be expected, decreased as T_m was approached [95].

Corresponding optical microscopy studies showed that the spherulite radii increased linearly with crystallisation time, i.e. spherulites had a constant rate of radial growth, indicating that the composition of the amorphous phase remained constant at the growth surface as crystallisation continued. Rates of radial growth increased with decreasing T_c over the limited temperature range investigated and were higher for pure PCL than for the blend; i.e. crystallisation was slower in the presence of PVC, a result consistent with the scattering data described in the preceding paragraph.

Kinetic data from scattering experiments could be analysed in terms of the Avrami equation (Eq. 26). This equation describes the development of crystallinity (or other separate phase) with time, at a constant temperature, in terms of a rate coefficient k and an exponent n; X_t is the fraction of crystallisable material which has crystallised at time t. The value of n reflects a combination of the kinetics of nucleation of the second phase and the number of dimensions in which crystallinity (or second phase) develops. A value of n=3 is consistent with a constant number of growing nuclei and growth in three dimensions, i.e. the kinetics of a constant number of expanding spheres. The Avrami equation allows for impinging growing phases and departures from the Avrami equation, at long times, are often explained in terms of secondary crystallisation within the already-formed spherulites:

$$X_t = 1 - \exp(-kt^n) \tag{26}$$

Values of the exponent n and rate coefficient k were estimated; values of $\tau_{1/2}$ (the times taken to reach 50% crystallinity) estimated from the analysis were consistent with values determined directly from times for the scattering intensity to reach $I_\infty/2$. Values of n were about 2 (about 2.2–2.4 for pure PCL and 1.7–1.9 for the blend) and are less than the value of 3 expected for uniform spherical growth with instantaneous nucleation (sample thicknesses were not quoted); values cannot be interpreted immediately but the difference in n for PCL and for the blends probably reflects differences in nucleation mechanism. Kinetic data from scattering and optical microscopy were consistent.

In a later extension of their earlier study, Nojima et al. [51] analysed the ring-banded structure of the PCL spherulites in PCL/PVC blends (PCL $\overline{M}_n$=27,400, $\overline{M}_w/\overline{M}_n$=1.59; PVC $\overline{M}_w$=114,000, $\overline{M}_w/\overline{M}_n$=1.56) cast from THF; film thicknesses were about 0.03 mm. (Fig. 5). Observations confirmed the previous studies by Russell and Stein [9] and Khambatta et al. [16] that pure PCL spherulites did not give a banded structure; ring-banded structures were only observed in blends. For blends of different compositions, weight fractions of PCL from 0.7 to 0.99, banded spherulites were observed with wide regular bands at high PCL contents and narrower less-regular bands at lower PCL contents. Widths of the bands when plotted approached infinite width at 100% PCL (Fig. 18). The interpretation of the data was that as the PVC content in the blend increased the PLC fibrils in the spherulites were induced to twist during growth, to twist with a shorter period and also more irregularly as the PVC content increased.

The observation that spherulites were space-filling at high PCL contents means that the PVC in the blend must be contained within the spherulites (interfibrillar or interlamellar). At 50 wt %, spherulites were found to be not quite space-filling and the observation of a single T_g in the partially crystalline samples means that the PVC and residual PCL were in a single homogeneous amorphous phase within and between the spherulites.

In a careful, detailed study, Russell and Stein [9] (PCL-700, Table 1) prepared samples cast from THF, dried at 40 °C under vacuum, melt pressed at 90 °C, crystallised at 30 °C and stored at room temperature. They reported the variation in PCL melting point with composition and weight fraction of PCL which crystallised. They reported that the PCL (as received) was coloured and had an odour. They purified the polymer by dissolution and reprecipitation. The polymer was found to contain two fractions with different molecular weights which were separated and dried under vacuum at 50 °C (gel permeation chromatography gave the weight-average molecular weights as 23,000 and 45,000; the latter fraction was used in the studies described below). Using purified samples these workers found no heterogeneity of PVC between lamellae in the spherulites, unlike the observations of Khambatta et al. [16] which were undertaken in the same laboratory. Degrees of crystallinity in blends based on purified samples of PCL were found to be lower than in unpurified samples. For blends containing <50% PCL the densities of blends containing purified PCL were equal to or higher than those based on unpurified material; the results suggested a negative volume of mixing or better miscibility (see Sect. 7). Above 50 wt % PCL, densities were dominated by the densities of the crystalline component and densities of blends with purified and unpurified PCL were virtually identical. Spherulite radii in purified blends were larger. These data were all consistent with there hav-

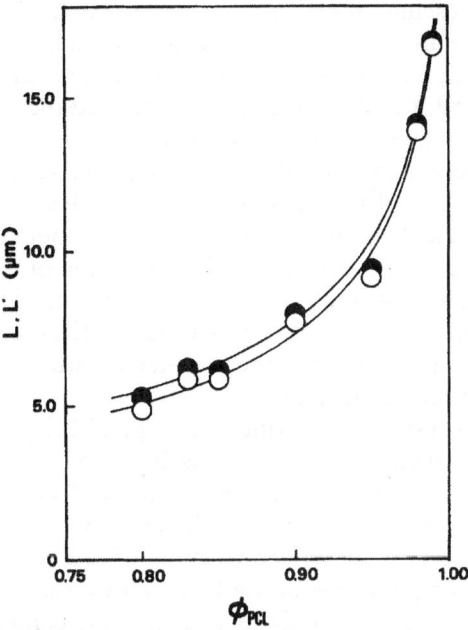

Fig. 18. Variations in average widths of the extinction rings with PCL content in ring-band-ed spherulites in PCL-PVC blends, calculated as (●) $L=\sum nl/\sum n$, (○) $L'=\sum nl^2/\sum nl$, where n, l are the numbers and widths of the bands, respectively; taken from [51]

ing been a lower content of nucleation sites in purified samples. That is, the re-sults suggest that an impurity present in commercial samples of PCL acted as a nucleating agent for PCL crystallisation. Thus, while intrinsically there was per-fect molecular miscibility in the amorphous component in pure PCL/PVC blends the same may not be true for samples prepared from commercial sam-ples; both the spherulite density and size may vary with polymer grade. This study by Russell and Stein appears to be the only study which refers to impurities and nucleating agents in commercial PCL [9]. An implication of this study is that miscibilities and blend properties could vary with the source of the PCL and the consequence of these observations must act as a reservation on the results of all studies of blends of PCL with other polymers.

Russell and Stein [9] also reported on the effects of thermal history on blend morphology. When samples were heated to just above the PCL melting point they found that recrystallisation resulted in the formation of rod-like crystals but if samples were heated above 90 °C recrystallisation gave spherulites. The implication of these data is that crystalline polymer, on melting, cannot diffuse into the amorphous material to form a uniform homogeneous phase at low tem-perature but can at high temperatures. This effect was not observed in earlier studies, possibly because of the influence of the low molecular weight fraction and nucleation in the blends.

Ong and Price presented thermograms for the melting of PCL/PVC blends cast from MEK [93]. Pure PCL exhibited a sharp endotherm with a peak at about 65 °C but with a weak tail extending to about 50 °C; the tail was attributed to a weak peak at about 57 °C. Addition of PVC broadened the melting endotherm. For blends with up to 25 wt % PVC the changes could be attributed to a strong growth of the component responsible for the low-temperature peak and a small decrease in temperature of the main peak to 63 °C. Further addition of PVC caused the minor peak to become the major peak, at about 50 wt % PVC, with further reductions in the temperature of the maximum of the higher tempera-ture peak, to about 58 °C at 60 wt % PVC; the maximum of the lower tempera-ture peak also shifted to about 54 °C at 60 wt % PVC. These changes were attrib-uted to changes in crystalline order; the high-temperature peak was not attrib-uted to the results of recrystallisation.

In their study of crystallisation kinetics, Ong and Price demonstrated con-stant rates of radial growth of spherulites (as discussed above) at various blend compositions [94], consistent with incorporation of PVC into the spherulites and retention of constant composition at the growth tips of the crystalline la-mellae. Rates of spherulite growth were quoted for a series of compositions and crystallisation temperatures in the temperature range 20–35 °C; rates of radial growth decreased with both increasing PVC content and increasing crystallisation temperature. Overall rates of crystallisation were analysed by Eq. (26) and values of the Avrami constants n and k were determined. Data were consistent with instantaneous heterogeneous nucleation and spherical spherulite growth.

Ajji and Renaud [96] investigated the mechanical properties of oriented PVC/PCL blends; weight-average molecular weights of PVC and PCL were 105,000 and 36,000, respectively. Films, cast from solution in tetrahydrofuran, were annealed under vacuum at 40 °C for 1 month and were subsequently drawn at various temperatures in the range 70–90 °C. For samples of PVC and of PCL-PVC blends with 25 wt % or 50 wt % PCL, they investigated T_gs and T_ms of the blends (Table 6) and the birefringences and mechanical properties of the orient-ed blends. They considered three possible deformation models for the interpre-tation of their data on blends low in PCL; namely pseudo-affine, affine and ex-tended affine, the last model assuming that orientation causes a disruption of an inherent physical network structure in the material and brings about a decreas-

Table 6. Glass-transition temperatures of blends of PCL with PVC; data taken from [96]

Wt % PVC	$T_g/°C$	$T_m/°C$
100	83	–
75	29	–
50	9	51
25	–30	53
0	–67	55

ing number of network segments. Experimental data gave good fits only to the extended affine model for PVC homopolymer and blends low in PCL. Their model assumed the existence of two types of network junction points with each type arising from different physical interactions.

None of the models considered could explain the behaviour of the blends containing 50 wt % PCL which contained some crystalline PCL as spherulites. At low elongations the spherulites did not break up but deformed. Spherulites started to break up when samples were extended beyond their yield points.

The mechanical properties of the unoriented blends were found to be complex. For oriented samples with 25 wt % PCL the modulus and stress at break increased with draw ratio; data for the 50/50 blend were more complex and not readily interpreted [96].

Polymer films are of interest both as permeable materials for gas transport and as barrier films. Gas permeability is therefore a parameter which can be usefully determined. The permeability of a gas depends on the product of its solubility in the polymer and its diffusion coefficient in the matrix; obviously both parameters may change with the composition of blends. Shur and Ranby investigated the permeabilities of oxygen and nitrogen through PCL-700/PVC($\overline{M}_w$= 74,000) blends [17]. The polymers were melt-mixed at 160 °C and formed into films 0.04 mm thick by pressing at 170 °C. Samples exhibited a single T_g but X-ray data showed the usual PCL crystallinity in samples containing 50 wt % or more PCL. The same pattern of behaviour was observed for the diffusion of both gases. Approximately linear variations in log(permeability) with PCL content were observed with a break point and maximum permeability at 30 wt % PCL. At low PCL contents the gas permeability increased with the PCL content as the materials became softer (plasticized by the polymer with the lower T_g) but at higher PCL contents the permeability decreased in the presence of crystalline PCL. Selectivity for gas separation appeared to be largely independent of PCL content. The data were consistent with true miscibility of PVC and PCL in the amorphous phase. Results are summarised in Fig. 19.

Pingping et al. examined the viscosity behaviour in PCL/PVC blends (50:50 w/w) in dilute solutions (<0.02 g cm^{-3}) in several solvents – 1,2-dichloroethane (DCE), N,N-dimethyl formamide (DMF) and tetrahydrofuran (THF) [97]. In each case the intrinsic viscosity (the limiting values of η_{sp}/c at zero concentration) of the PCL used was about 0.2 dl g^{-1} and values of η_{sp}/c varied little with concentration. Intrinsic viscosities of PVC in DCE, THF and DMF were 1.7, 1.0 and 0.76 dl g^{-1}, respectively; in THF and DMF values of η_{sp}/c increased with PVC concentration but in DCE decreased with increasing polymer concentration. In the three solvents the blends had intermediate viscosity behaviour; variations of η_{sp}/c with concentration were slight and in DCE the dependence was non-linear. The authors defined ideal solution behaviour for a binary mixture of polymers A and B as one for which the relation

$$(\eta_{sp}/c)_{c \to 0, blend} = [\eta]_A (c_A/c)_{c \to 0} + [\eta]_B (c_B/c)_{c \to 0}$$

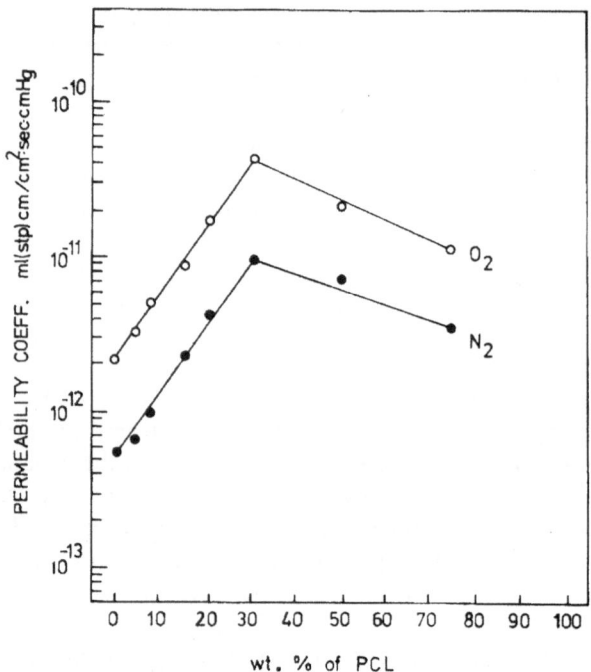

Fig. 19. Variations in gas permeability at 25° C in PCL-PVC blends with PCL content for (○) oxygen and (●) nitrogen permeation; taken from [17]

where $[\eta]_I$, c_I are the intrinsic viscosity and concentration of species I, respectively, holds true. For the solutions investigated the intrinsic viscosities of the mixtures in THF and DMF were in reasonable agreement with this relation but the calculated value of intrinsic viscosity of solutions in DCE was lower than that calculated from the intrinsic viscosities of the individual components. The intrinsic viscosity in DMF was about 20% greater than that calculated, an effect attributed to attractive interactions between PCL and PVC increasing the hydrodynamic volume overall. The low value of intrinsic viscosity observed in DCE was attributed to attractions between the two polymers reducing polymer-solvent contacts between PVC and DCE in which PVC has a high intrinsic viscosity and hydrodynamic volume as a result of strong interactions between them as a result of their similar chemical structures. This study has limited relevance to the solvent dependence of blend morphology on casting solvents because the data were determined at very low polymer concentrations where polymer-polymer contacts are limited and systems are very mobile.

9
Blends with Chlorinated Polyethylenes

PCL is but one polyester of a series of related materials with different proportions of methylene and ester groups; i.e. with different values of n in 2. Both Riedl and Prud'homme [72] and Ziska et al. [98] agree that, of the aliphatic polyesters, PVC is only miscible with PCL. On the other hand, PVC can be considered as a limiting and specific structure in the family of chlorinated polyethylenes, i.e. one which contains 56.3 wt % chlorine and in which the chlorine atoms are distributed in a regular manner, one on every other carbon atom, along the polymer chain. Chlorination of polyethylene produces a series of materials with different chlorine contents in which the chlorine atoms are distributed randomly along the polyethylene chain; there may be chlorine atoms on adjacent carbon atoms and more than one chlorine atom per carbon atom. In considering the series of blends of PCL with chlorinated polyethylenes we recognise different compositions of the chlorinated polyethylenes and these are identified as CPE-x where x is the wt % of chlorine in the chlorinated polyethylene.

Crystallisation of PCL ($\overline{M}_n$=14,000, $\overline{M}_w/\overline{M}_n$=1.6) in its blends with chlorinated polyethylene containing 56.3 wt % of chlorine (i.e. with CPE-56.3, Table 7) was studied by Defieuw et al. [99]; CPE-56.3 has the same overall chlorine content as PVC but the chlorine atoms in CPE-56.3 are distributed randomly along the chain. The glass-transition temperature of CPE-56.3 is 47 °C compared with 85 °C for PVC. As in PVC blends, PCL crystallises to give space-filling spherulites and rates of radial growth for isothermal crystallisation are constant. Thus there is no liquid-liquid phase separation in the blends (blends are miscible above T_m of PCL) and the CPE is located intraspherulitically (cf. previous discussion of PVC blends). Rates of radial growth of spherulites, in blends rich or dilute in PCL, were found to be higher than for PVC blends. The higher rates of crystallisation in CPE-56.3 blends were attributed to the lower T_g of CPE-56.3, compared with PVC, and the consequent higher molecular mobility, allowing more rapid diffusion and organisation of PCL chains. Notably the spherulites formed did not exhibit the banded structure found with PVC blends; the texture seen in photomicrographs is strongly fibrillar (Fig. 20).

SAXS experiments [99] showed a decrease in long-spacing in PCL spherulites in CPE-56.3 blends (in direct contrast to the increase in long spacing observed in PVC blends) after isothermal crystallisation. While the increase in long spacing in blends with PVC was attributed to interlamellar segregation, the observed decrease in chlorinated polyethylene blends was attributed to interfibrillar segregation of residual amorphous components in spherulites of different structure, despite the appearance of a single glass-transition temperature.

Defieuw et al. [100] also investigated blends of the same PCL with CPE of different Cl contents with chlorine atoms distributed randomly on the polymer backbone; molecular weight data are given in Table 7. Although lightly chlorinated polyethylenes are partially crystalline, CPE-30.1 is amorphous and only amorphous CPEs were studied. Samples were prepared by co-precipitation of

Table 7. Characterisation data for chlorinated polyethylenes used in blends with PCL [99]

Chlorine content (wt %)	$\overline{M}_n$	$\overline{M}_w / \overline{M}_n$	$T_g/°C$
35.6	59,000	3.35	−10
42.1	50,000	3.85	1.5
49.1	53,000	4.64	18.5

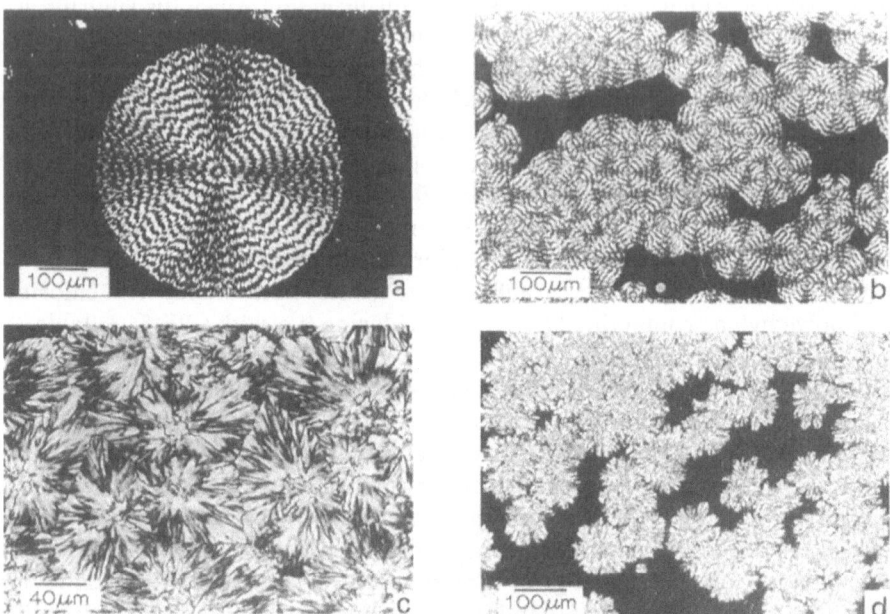

Fig. 20a–d. Optical micrographs showing the distinction between ring banded spherulites in PCL-PVC blends and non-banded structures in blends of PCL with CPE-56.3; taken from [99]

the components from homogeneous solution and precipitated polymers were then compression moulded at 120 °C. The miscibility of PCL with chlorinated polyethylenes varied with both the extent of chlorination and PCL content. PCL was reported to be miscible with CPE-49.1 in all proportions at temperatures up to at least 200 °C, although PCL crystallised from PCL-rich blends. CPE-42.1 was miscible over a limited range of temperatures and compositions and exhibited LCST behaviour (Fig. 21). CPE miscibility with PCL decreased still further at lower chlorine contents.

To obtain totally amorphous samples, blends with, say, CPE-42.1 were heated to 100 °C, above the melting point of PCL but below any LCST, and quenched in liquid nitrogen. Even so, blends with 80 wt % or more PCL were not totally amorphous because the extremely strong tendency of PCL to crystallise could

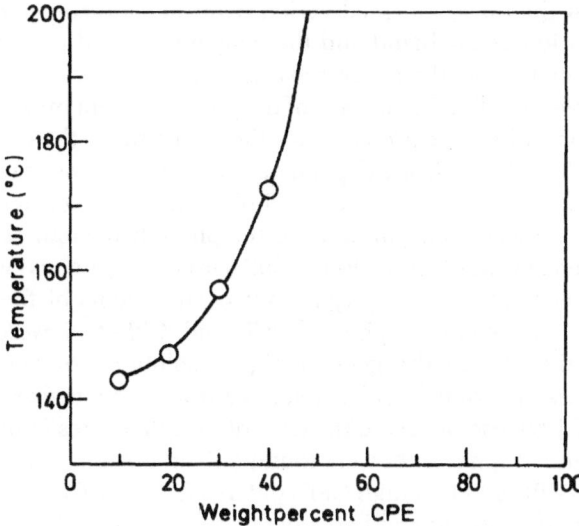

Fig. 21. Cloud-point curve for blends of PCL with CPE-42.1, showing LCST-type behaviour; taken from [100]

not be totally suppressed. Amorphous materials gave reasonable agreement between experimental T_gs and those calculated from Eq. (23) (using $T_g \sim -63$ °C for PCL); estimated T_gs for pure amorphous PCL obtained by other workers were about −70 °C (see Sects. 2 and 8) and use of this value would also give a reasonable fit to Eq. (23). For samples containing at least 50 wt % of PCL, thermograms obtained by differential scanning calorimetry indicated that the PCL crystallised above T_g of the mixture. Amorphous 50/50 blends of PCL with CPE-49.1 exhibited a single T_g by dynamic mechanical analysis but after crystallisation of the PCL the T_g was raised, reflecting an increased content of CPE-49.1 in the residual amorphous phase.

Growth rates of PCL spherulites in blends with CPEs were found to depend on both CPE and Cl contents [100]. The growth rate at T_c=45 °C decreased by about a factor of ten on adding 20 wt % CPE-42.1 (T_g=1.5 °C) to PCL. In 50:50 (w/w) blends the growth rate in the presence of CPE-49.1 (T_g=18 °C) was about a factor of four lower than for CPE-42.1. Decreased growth rates were partly a consequence of dilution of PCL and partly because of the increased viscosity in the amorphous phase on adding CPE with a higher T_g. It was also noted that growth rates of spherulites in the presence of 20 wt % CPE-42.1 were slightly higher if the blend had been heated to a temperature above the LCST for 5 min than if kept below the LCST. This effect was caused by partially phase-separated systems not being instantaneously homogenised on cooling below the LCST; one phase had a higher PCL content than the overall sample.

In blends with low-chlorine-content CPEs (<42.1 wt %) the LCST decreased sharply and partially miscible systems were common [100]. Dimensions and compositions of the phases formed varied with the CPE chlorine content, the overall composition of the blend and the temperature and time for which the blends were heated above the phase-separation temperature. Such systems are obviously complex and the PCL in each of the phases present may phase separate by crystallisation; spherulite growth rates then depend on the CPE contents in each phase and T_gs of the phases depend on the compositions of residual amorphous components. A phase-separated PCL/CPE-35.6 50:50 (w/w) blend exhibited only one T_g. Observed asymmetry in the phase behaviour (PCL/CPE-35.6 10:90 is homogeneous but PCL/CPE-35.6 90:10 is heterogeneous) was, at least in part, a consequence of the disparity in molecular weights of the components (weight-average molecular weights of PCL and CPE-35.6 were 22,500 and 198,000, respectively) and the general higher solubility of small molecular weight substances in high molecular weight substances than vice versa.

Defieuw et al. [99] also observed the textures of spherulites found in different blends. Ring-banded spherulites were found only in systems with low CPE contents (~10 wt % CPE-49.1). At high CPE contents (20 wt % CPE-49.1) the spherulites were more disordered and had a dendritic texture. At high PCL contents (at least 70 wt %) spherulites were found to be space-filling and the CPE was necessarily contained in the amorphous regions of the spherulites. For higher CPE contents (of CPE-49.1) spherulites were found to be separated by a CPE-rich amorphous phase. Details were stated to depend on the detailed experimental procedures, e.g. on the crystallisation temperatures which influenced rates of crystallisation and of molecular diffusion.

The melting behaviour of blends with chlorinated polyethylenes (and PVC) also depends on several factors. These systems often exhibit two melting endotherms when investigated by DSC and T_m was taken to correspond to the high-temperature endotherm. For blends in which PCL had crystallised from a single amorphous phase the melting temperature (T_m) varied with the CPE content [100].

Values of T_m decreased with increasing CPE content, from 59.5 °C for 100% PCL to 52.3 °C for 50 wt % CPE-49.1. For CPE-42.1 blends, variations in T_m were less marked (Fig. 22). Defieuw et al. [100] attributed the differences in behaviour to differences in lamellar thickness, resulting from differences in viscosity and changes in the interaction parameter. These observations were consistent with SAXS data which identified a smaller value for the long-spacing in CPE-49.1 blends, decreasing from 13 nm in pure PCL to 10 nm with 30 wt % CPE-49.1.

T_ms of blends which had phase separated prior to crystallisation were little affected by the CPE because the CPE concentration in the PCL-rich phase formed was low; high molecular weight CPE had little solubility in low-molecular-weight PCL. T_m was 59.1 °C in a blend containing 30 wt % CPE-35.6 and even a 10/90 blend with CPE-35.6 exhibited a small PCL melting endotherm.

A general feature of the melting behaviour of blends which are miscible in the melt is a pronounced, small, low-temperature melting endotherm observed in DSC experiments, which might occur 10 °C or more below the main endotherm, especially at low scan rates. Melting endotherms from phase-separated blends, e.g. CPE-35.6 blends, may show broad multiple endotherms. In the latter case

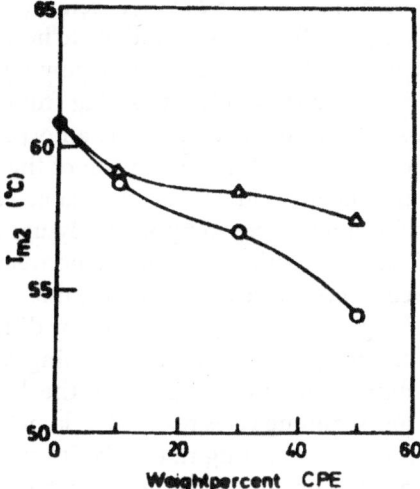

Fig. 22. Variations in PCL melting point (T_{m2}) in blends with (O) CPE-42.1 and (Δ)CPE-49.1; taken from [100]

the spherulites are not space-filling and there is crystalline material in both the highly-crystalline PCL-rich and the CPE-rich phases.

Defieuw et al. [100] also investigated the influence of crystallisation conditions on the melting behaviour for CPE-49.1 blends (40 wt % CPE), which were homogeneous prior to crystallisation. After crystallising at 25 °C the higher melting peak (in DSC thermograms) which developed moved to a slightly lower temperature, and attained a constant area after about 180 min. The lower temperature endotherm continued to increase in size and was attributed to additional crystallisation. This secondary crystallisation occurred within the space-filling spherulites and must have arisen from further crystallisation of PCL from the residual CPE-rich amorphous residue.

The phase behaviour of copolymers of ε-caprolactone and ε-caprolactam blended with PVC, chlorinated polyethylene and chlorinated PVC have been investigated. While PCL itself is miscible with these chlorinated polymers, the influence of replacing ester by amide linkages was investigated. While polycaprolactam is immiscible with these chlorinated polymers, it was demonstrated that a high level of caprolactam units in the copolymer was necessary to induce immiscibility of these copolymers with the chlorinated polymers [101].

10
Blends of Vinylidene Chloride Copolymers (Saran)

A series of vinylidene copolymers are sold commercially as Saran. These are random copolymers of vinylidene chloride with vinyl chloride (88 mol % vinylidene chloride, $\overline{M}_w$=100 kg mol^{-1}, T_g=7 °C) 4, acrylonitrile (>80% vinylidene chloride, $\overline{M}_w$=135 kg mol^{-1}, T$_g$=56 °C) **5** or vinyl acetate (>80% vinylidene chloride, $\overline{M}_w$=82 kg mol^{-1}, T$_g$=0 °C) **6**. Zhang and Prud'homme investigated the miscibility of PCL ($\overline{M}_w$=20,000) with each of these polymers [74]. They demonstrated that each blend, prepared by solvent casting from tetrahydrofuran, exhibited a single composition-dependent T$_g$ for the amorphous component. The vinylidene chloride copolymers, with the exception of the acrylonitrile copolymer, were crystalline and blends with the crystallisable copolymers exhibited crystal melting points for at least one component. By use of Eq. (10) they estimated values of the binary interaction parameters between PCL and the Saran polymers; Hoffman-Weeks plots were used to determine values of the equilibrium melting points. Melting points were determined by differential scanning calorimetry and were found to increase with the scanning rate if that rate was in excess of about 5 °C min^{-1} and more rapidly if the heating rate exceeded 30 °C min^{-1}. However, low scanning rates resulted in lamellar thickening during heating and, as a compromise, a heating rate of 20 °C min^{-1} was used in these studies. The values of the interaction parameters, which were found to vary with composition, were similar for vinylidene chloride-vinyl chloride and vinylidene chloride-vinyl acetate copolymers; values were determined from the melting point depressions of the major component in each case (Table 3). For vinylidene chloride-acrylonitrile copolymers values of the interaction parameter (all negative) could only be determined for blends rich in PCL and, apart from a very nega-

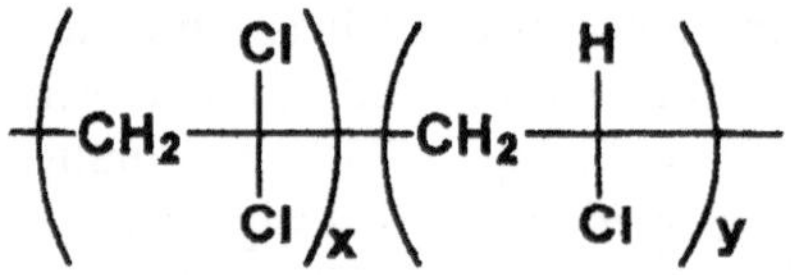

Structure 4

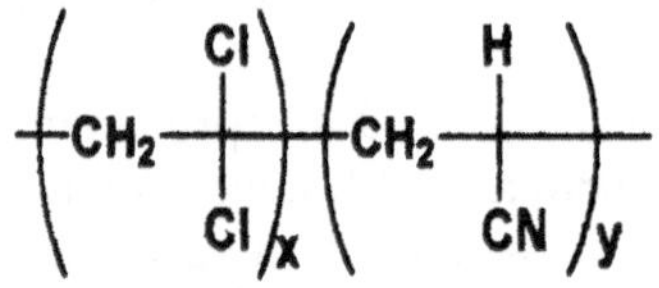

Structure 5

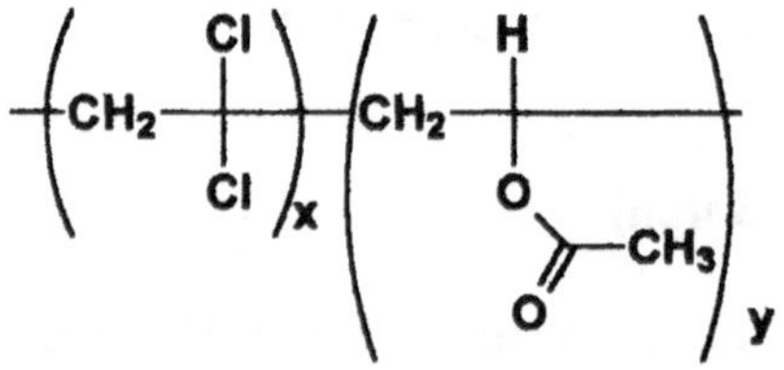

Structure 6

tive value recorded for 95 wt % PCL, the values observed were virtually compo-sition independent. The values fell on a smooth curve even though values in PCL-rich samples were determined from PCL melting points (~60 °C) and in vi-nylidene chloride-rich blends from the copolymer melting points (~180 °C), in-dicating little temperature variation in interaction parameter (see Fig. 13) [74].

11
Blends with Styrene-Acrylonitrile Copolymers

Blends with styrene-acrylonitrile copolymer (SAN) 7 represent an important se-ries of systems in which a crystallisable polymer, PCL, is blended with an amor-phous (co)polymer. The overall system, with its inherent variations, is complex and provides a good example of situations which can arise generally. These blends have been the subject of a major series of studies which raise many issues which must have a relevance to blends of PCL with other polymers but which have not been addressed in studies of other systems.

PCL is reported to be miscible with random (statistical) SAN copolymers but only over a certain range of copolymer compositions. Thus Chiu and Smith ex-amined a series of blends with different PCL/SAN compositions using various SAN copolymers with AN contents in the range 0–36 wt % [19, 20]. (SAN-n is a SAN copolymer with n wt % AN). On the basis of glass-transition-temperature measurements, and taking a single (intermediate) T_g to denote compatibility, they concluded that copolymers with AN contents of 8–28 wt % are compatible with PCL in all proportions [20], except for SAN-28 blends containing more than 70 wt % PCL [19]. This is an example of a system in which the homopolymers of the components of the copolymer, PS and PAN, and the other constituent of the blend are all immiscible with each other in pairs. That is, the interaction param-eters between all structural units are unfavourable for mixing but a "miscibility window" exists where copolymers of certain compositions are miscible with the third polymer. This result implies that the interaction parameters, in this case, between PCL and PS and between PCL and PAN are less unfavourable to mixing than are the interactions between S and AN units in the copolymer. Svoboda et al. reported negative enthalpies of mixing for PCL with copolymers rich in S units (see Sect. 7 on interactions) [77].

The above statement on compatibility in all proportions, however, does not imply total miscibility in all compatible mixtures under all conditions. Chiu and

Structure 7

Smith used blends cast from dichloromethane, employing PCL-700 (Table 1) and SAN copolymers with $\overline{M}_n$ in the range 200,000–270,000 and of uniform composition (composition variation in the copolymer was avoided under the conditions used for copolymer synthesis) [20]. Blends, transparent at room temperature, were only observed for SAN contents >75 wt % and AN contents in the range quoted above. At these compositions there was no PCL crystallinity and all polymer was contained in a single homogeneous amorphous phase. For SAN contents <75 wt %, in the same range of S/AN ratios, translucent or opaque blends were observed which became transparent on heating above the PCL T_m. These blends contained smaller or larger amounts of crystalline PCL and the single T_gs refer to the SAN and residual amorphous PCL. Above T_m of PCL, where the crystalline PCL melted, the blends were miscible and PCL recrystallised on cooling; the times required for opacity to develop as a result of crystallisation at 25 °C increased with increasing SAN content. (At higher SAN contents the driving force for crystallisation would be less and the viscosity of the mixture higher because of the higher T_g of SAN than of PCL; both factors would reduce the rate of crystallisation.)

Just outside the stated compatibility range the blends were opaque at room temperature and when heated above the T_m of PCL, indicating liquid-liquid phase separation in the melt as well as phase separation through crystallisation of PCL. Further outside the compatibility range there was gross phase separation and effective total separation of the components with poor adhesion between phases.

The thermal transition behaviour of the blends was consistent with the above observations. For the "compatible" blends (with SAN-8 to SAN-24) the dynamic mechanical properties showed, at high SAN contents where there is no PCL crystallinity, a single shifted T_g consistent with plasticization of SAN by admixing with PCL. Pure SAN copolymers had T_gs from 116 °C (SAN-8) to 124 °C (SAN-28). At higher PCL contents, where there is PCL crystallinity, the shifted T_g overlapped with PCL melting to give a complex relaxation peak; the β-relaxation of PCL was seen at –120 °C [20].

SAN-28 blends behaved similarly for compatible blends containing up to 50 wt % (or more) PCL, although differences were noted between the relaxation behaviours of "as cast" and quenched samples (the latter samples were quenched to low temperature after heating above the PCL T_m). For incompatible blends containing 70% or more PCL the relaxation behaviour was a combination of α-relaxation peaks of the two component polymers and the β-relaxation of PCL [19]. Blends of SAN-30 (70 wt % PCL) simply exhibited a combination of α- and β-relaxations of the two polymers [20].

Glass-transition data determined by DSC were similar to but slightly different from dynamic mechanical data. As-cast samples at high SAN contents and with no PCL crystallinity showed very similar single, composition-dependent T_gs by both techniques. At higher PCL contents (30–70%, approximately), as-cast samples showed a constant T_g (~50 °C) by dynamic mechanical analysis; these data describe blends with combinations of different proportions of crystalline PCL and a mixed amorphous phase of constant composition. At lower PCL contents T_g decreased as the SAN content of the amorphous phase decreased and T_g approached that of PCL [19].

T_gs determined by DSC of samples quenched from 167 °C, to inhibit PCL crystallisation, showed similar results to those from dynamic mechanical analysis from high SAN content samples where PCL does not crystallise. At lower SAN contents DSC and dynamic mechanical data showed slight differences, the latter exhibiting slightly lower T_gs than by DSC for which the T_gs decreased monotonically and consistent with the Gordon-Taylor equation (Eq. 22) (assuming a value of T_g for PCL of −83 °C and a value of k=0.576 which was determined from blends of SAN-24 [20]) to about 36% PCL after which T_gs were higher than the Gordon-Taylor prediction but they probably reflected the presence of some crystallinity in the PCL. These data implied some system dependence of T_g.

Blends which showed gross incompatibility, e.g. SAN-30 and SAN-33 blends, exhibited two T_gs consistent with total phase separation.

Cloud-point curves for the several blends showed a consistent pattern of behaviour [20]. Blends of copolymers containing 24 wt % or less AN exhibited homogeneity, i.e. true miscibility, over the whole composition range above T_m of crystalline PCL until shallow cloud-point curves, with LCST behaviour associated with liquid-liquid phase separation, were reached at higher temperatures. For blends well within the compatible range of AN contents the LCST was about 240 °C at 20 wt % SAN, approximately. At the boundaries of the "compatibility" range the cloud points were lower (about 180 °C for SAN-8) or unobservable (below T_m for PCL) for SAN-28. This cloud-point curve was noted to be at lower temperatures than that reported by MacMaster [39] for the same system where the cloud-point curve was observable over the whole composition range with a critical temperature of 87 °C at a volume fraction of SAN of 0.17. The difference in behaviour in the two studies is consistent with the use by Chiu and Smith [20] of higher molecular weight SAN copolymers which reduces miscibility and temperatures for phase separation. Cloud points changed very rapidly with AN content in the SAN copolymer for blends of a given overall PCL/SAN ratio (Fig. 23).

Several of the preceding experiments point out the complexity of the PCL/SAN system where detailed results depend on several parameters including, in addition to the molecular weights of the constituent polymers, the composition of the SAN copolymer. Suggestions of further complexities are hinted at in small changes in the differences between T_gs obtained by DSC and dynamic mechanical analysis with composition of the blends.

In an ancillary experiment, Chiu and Smith [20] showed that samples, miscible on solvent casting (30% PCL), quenched after heating above the cloud point for a very short time showed evidence of PCL crystallinity and that the higher the temperature to which they were heated prior to quenching the greater the PCL crystalline content. Simply, this is a consequence of quenching preventing phase-separated PCL redissolving in the SAN phase during the brief period while above its T_g but allowing PCL to crystallise within the mobile PCL phase.

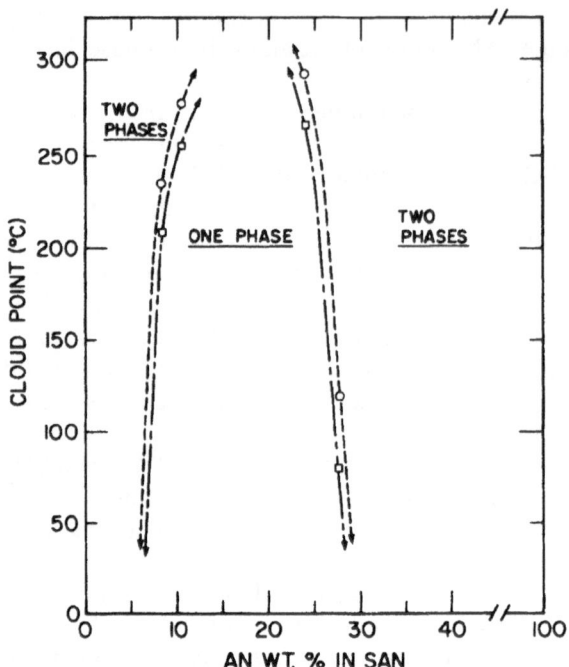

Fig. 23. Cloud point curves in PCL/SAN blends as functions of SAN composition for (O) SAN/PCL 70:30 (w/w), (□) SAN/PCL 50:50 (w/w); taken from [20]

A similar experiment with slow cooling allowed PCL to redissolve on the extended time scale and no PCL crystallinity was induced.

In related experiments with SAN-24, Chiu and Smith [19] analysed blends prepared by solvent casting, followed by drying under vacuum and subsequent compression moulding at 130 °C and cooling at 1.5 °C min⁻¹. Samples were subsequently annealed at 45 °C for three days and at 25 °C for one week. Such samples were analysed by DSC (heating rate 20 °C min⁻¹), cooled in the DSC (20 °C min⁻¹) and reanalysed on a second scan. The first scan on annealed samples gave T_m typically 20 °C higher than on a second scan and levels of crystallinity observed in the annealed samples were also higher, typically 15–20%. Samples cooled in the DSC were considered to have poorer crystallites indicating an influence of properties on thermal history of the samples. A comprehensive phase diagram was obtained (Fig. 24) together with plots of T_g variations (Fig. 25) [19].

The non-equilibrium nature of PCL/SAN blends and the difficulties in determining definitive properties of these blends was highlighted by an extensive series of investigations by Rim and Runt [102, 103]. These workers reported the effects of conditions of sample preparation in solvent-cast and melt-blended materials on the crystallinity of the blends [103] and on their melting behaviour [102].

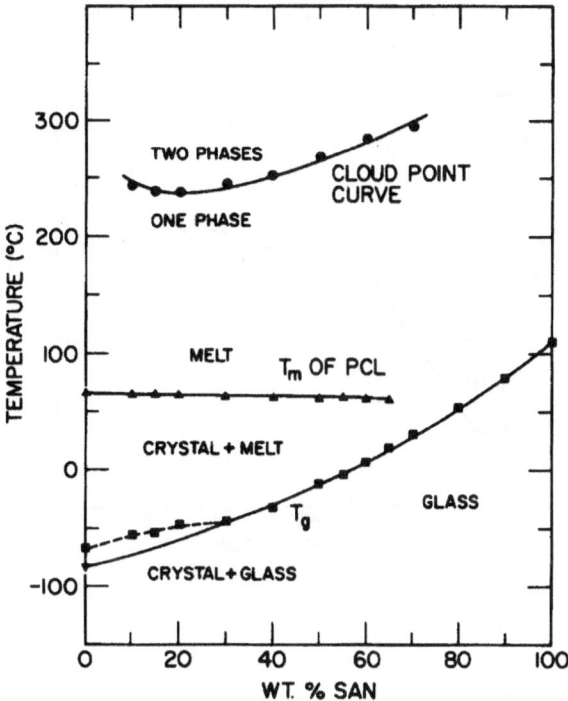

Fig. 24. Phase diagram for PCL/SAN blends; taken from [19]

Rim and Runt investigated the melting behaviour of blends of PCL ($\overline{M}_w$= 40,000) blends with 10–90 wt % SAN-24 (molecular weight not quoted) prepared by solvent casting [102]. Samples were melt crystallised by heating in the DSC instrument to 87 °C, quenched and reheated to room temperature, and their overall crystallinities were determined by DSC. Crystallinities decreased with increasing SAN content to about zero at 70 wt % SAN-24. When overall crystallinities were converted to PCL crystallinities the percentage of PCL crystallised was approximately constant at 60–70% from pure PCL to about 60 wt % SAN, below which the PCL crystallinity also decreased. The data were analysed in terms of T_gs of the amorphous blends. Samples with T_gs lower than the crystallisation temperature (room temperature) allowed crystallisation to occur. Samples with high SAN content and T_gs higher than room temperature prevented crystallisation of the PCL. At intermediate SAN contents it was considered that PCL could crystallise until the composition of the residual amorphous phase reached the composition where its T_g equalled the crystallisation temperature, approximately, when crystallisation ceased. Experimental data were generally consistent with this model.

Rim and Runt discussed melting behaviour in blends in general. They pointed out that T_m depends on thermodynamic factors, melting-point depression and

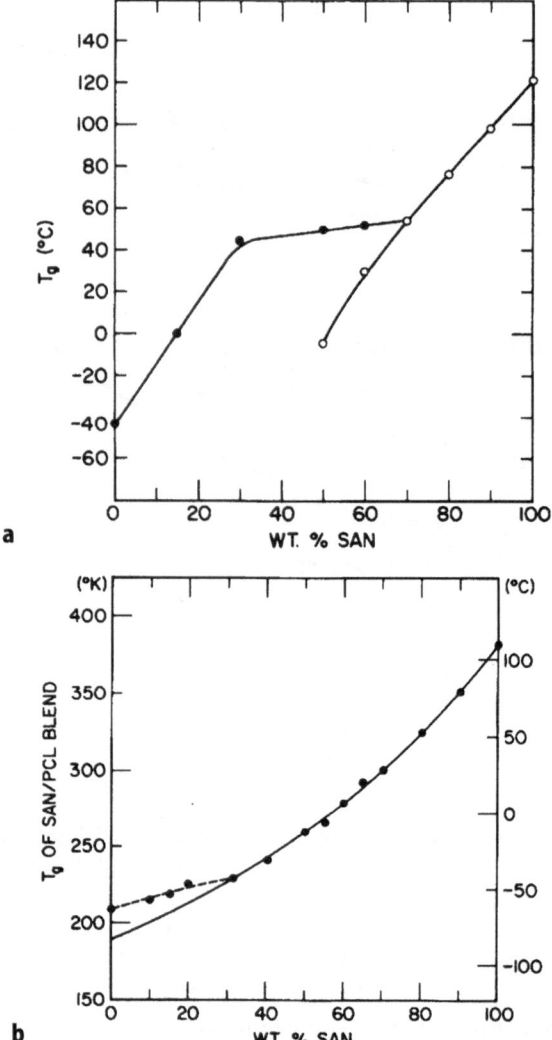

Fig. 25a,b. Variations in glass-transition temperatures in PCL/SAN blends: **a** for quenched samples (O) without PCL crystallinity and (●) with PCL crystallinity determined by dynamic mechanical analysis at 110 Hz; **b** determined by DSC with *solid line* calculated from the Gordon Taylor equation with $k=0.576$ and T_gs for PCL and SAN 190 K and 384 K respectively; data taken from [19]

lamellar thickness, and that other factors might also influence the observed value of T_m. For example, some crystallites might melt at low temperature and recrystallise while heating to give a high observed T_m. Depending on the rates of heating and recrystallisation a single, high-temperature melting endotherm or multiple melting peaks might be observed. In general any observed T_m may not be that of the as-formed crystals. The equilibrium melting point might be influenced by polymer-polymer interactions (see PVC blends, Sect. 8). Crystallite perfection may vary and this factor can also influence T_m.

Solution-cast samples exhibited only a single composition-dependent T_m; T_m was highest for pure PCL and there was a variation over about 30 °C. This variation might be due to thermodynamic or non-thermodynamic effects, as discussed above.

Melt crystallised samples, in general, showed two melting endotherms. The lower melting peak increased in relative size with increasing SAN content; the temperature difference between the two peaks was about 5 °C except at high PCL contents when the lower temperature peak was shifted to lower temperatures (by 10 °C, approximately). For samples containing 70 wt % PCL the relative sizes and positions of the endotherms varied with the heating rate. Observed T_ms generally varied smoothly with composition but the overall behaviour observed was complex and T_ms both decreased and then increased with increasing heating rate [102]. The authors discussed the overall results in terms of a mechanism in which an increase in molecular mobility in the amorphous state allowed molten material to crystallise or recrystallise. They were able to predict the general trends in behaviour observed. All these data point to the non-equilibrium nature of the samples and melting points and the dependence of sample properties on thermal history.

In a general discussion of melting behaviour, Rim and Runt [104] reiterated that it is virtually impossible to determine the true experimental melting points of "as formed" crystals in PCL/SAN blends due to reorganisation during heating; the problem is at least partially the limited thermal conductivity of the sample and difficulty in achieving sufficiently fast heating rates. It follows that the same thermal conductivity will restrict rates of cooling and will determine the thermal history of the interiors of all thick samples.

The mechanical properties of blends in which one component is crystallisable have generally not been subjected to such extensive investigation as wholly amorphous blends. Although amorphous blends might be expected to exhibit simple, general relationships between the properties of the components and of the blends, variations, such as enhancement of properties, might be associated with reduced volumes, due to polymer-polymer interactions, or the development of some specific morphology. Rim and Runt studied the mechanical properties of solution-cast and melt-crystallised sample of PCL/SAN-24 blends [105]. Solutions of the polymers were mixed, solvent was removed under vacuum, and samples were then aged at room temperature for two months; ageing could involve crystallisation or densification of a glass if the sample's T_g was below room temperature. Melt-crystallised samples were heated to 87 °C for 2 min, quenched and then aged at room temperature for 2 months.

Initial tensile moduli of solvent-cast samples decreased with increasing PCL content. At low PCL contents this effect was due to the addition of PCL, with a low T_g, to SAN-24 to give an amorphous material of lower T_g. The characteristics of the stress-strain curve changed from brittle fracture for pure SAN-24 to the

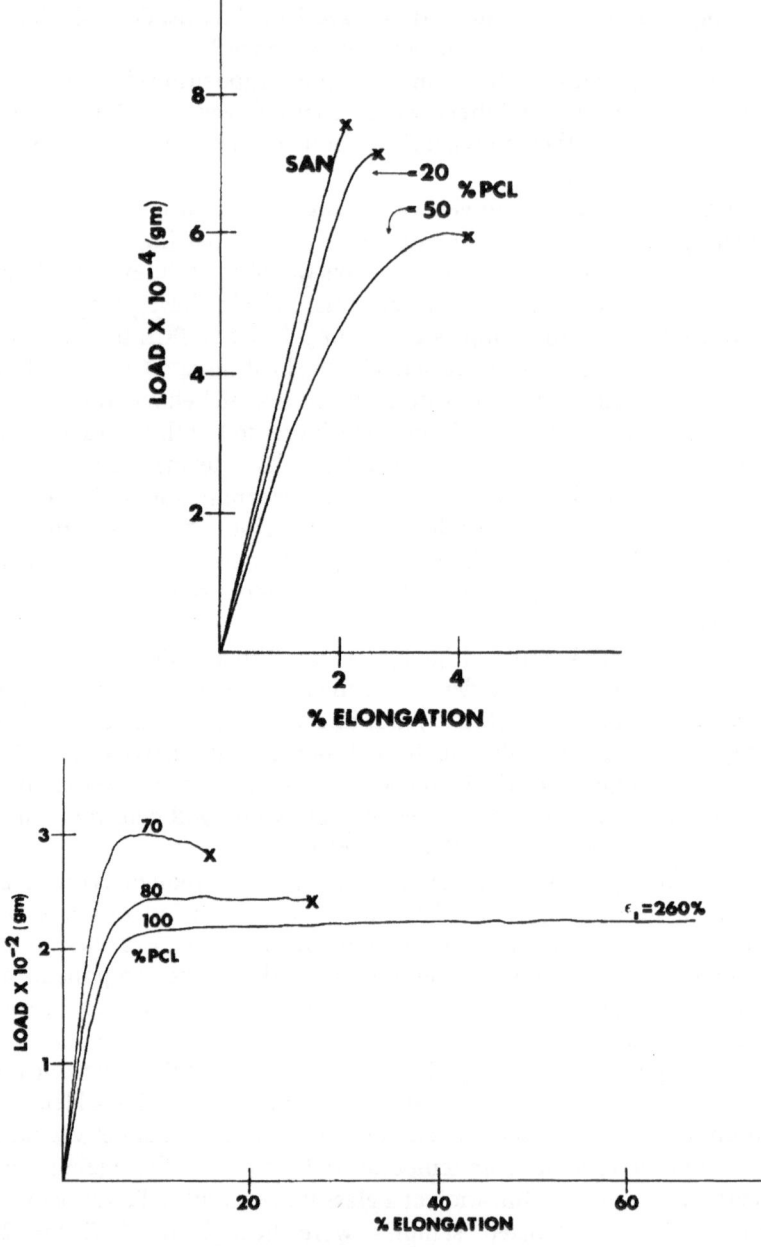

Fig. 26. Load elongation curves for PCL/SAN blends, solution cast, with different PCL contents; taken from [105]

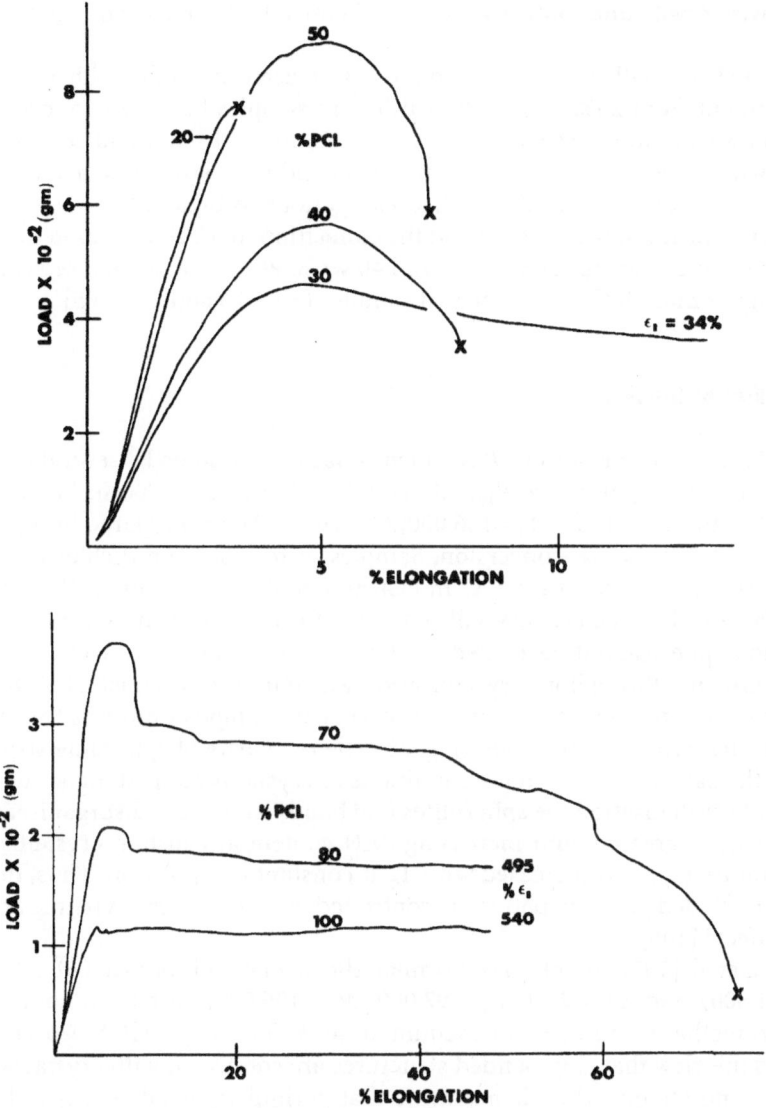

Fig. 27. Load elongation curves for PCL/SAN blends, melt crystallised, with different PCL contents; taken from [105]

appearance of a small yield at 50 wt % PCL with increasing yields and loss of modulus and ultimate strength as the PCL content increased to 100%. There was a small deviation (increase in modulus) from the general trend when PCL contents reached about 30 wt % as the PCL started to crystallise; the crystallites reinforced the matrix and the loss of PCL in the amorphous phase increased T_g of the remaining amorphous material. Rim and Runt stated that variations in properties with crystalline content were overwhelmed by the effects of variations in T_g.

For melt-crystallised samples, moduli decreased markedly with increasing PCL content from 20% PCL to 30% PCL and samples became very extensible (34% extension to break compared with ~2% for PCL) and had reduced T_gs. Then, with more PCL, moduli rose markedly and yield strengths increased, but samples with up to 50 wt % PCL had elongations to break of about 7%. With more PCL the moduli decreased and the elongation to break increased in general. In most cases, except at 30 wt % and 40 wt % PCL, melt crystallised samples have higher moduli than solvent-cast samples (Figs. 26 and 27) [105].

11.1
Crystallisation Studies

Crystallisation of various PCL/SAN blends has been studied by several groups.

Kressler et al. [106] investigated crystallisation of PCL-700 in blends with SAN-19.2; the SAN-19.2 ($\overline{M}_w$=166,000, $\overline{M}_w/\overline{M}_n$=2.2) was prepared in copolymerisations taken to low conversion. Samples were cast from dichloromethane and dried under vacuum at 60 °C. In PCL and in blends containing 10 wt % and 40 wt % SAN-19.2 the PCL crystallised in the form of spherulites; crystal nuclei were non-spherical but developed a spherical form on growth. The number of nuclei growing throughout crystallisation was found to be constant. A general feature of the spherulites formed at several blend compositions was the formation of ring-banded spherulites. (Fig. 28) The regularity of spherulite structure and of the bands varied with the conditions of crystallisation. At higher temperatures of crystallisation the spherulites and bands were more disorganised; ring periodicity decreased with increasing SAN content at constant crystallisation temperature (T_c) and increased with T_c at constant composition. Thus, the authors concluded, the ring pattern is controlled by both thermodynamic and kinetic effects [106].

Wang et al. [107] also observed similar effects in blends of PCL ($\overline{M}_w$=22,000, $\overline{M}_n$=11,300) and SAN-25 ($\overline{M}_w$=197,000, $\overline{M}_n$=106,000) cast from solution in dichloromethane and dried in vacuum at 40 °C for 3 days. These workers expressed the view that ring-banded structures are connected with crystallisation kinetics and attempted to demonstrate that periodicity is inherently linked to mobility of chain segments. They found no obvious relationship between rates of radial growth and periodicity (however, see Sect. 4); the former depends on rate of transport of crystallisable segments across the liquid-solid interface (related to T_g of the crystallising mixtures) and the free energy of critical nucleus formation on crystal surface (related to degree of undercooling). At a given temperature (of crystallisation) the overall segmental mobility is affected by the T_g of mixture. Mobilities will be modified by specific interactions between the crystallisable and non-crystallisable components. In this case, PCL mobility is reduced through the higher T_g of the amorphous SAN. Overall mobility, during crystallisation, can be represented by the difference between crystallisation tem-

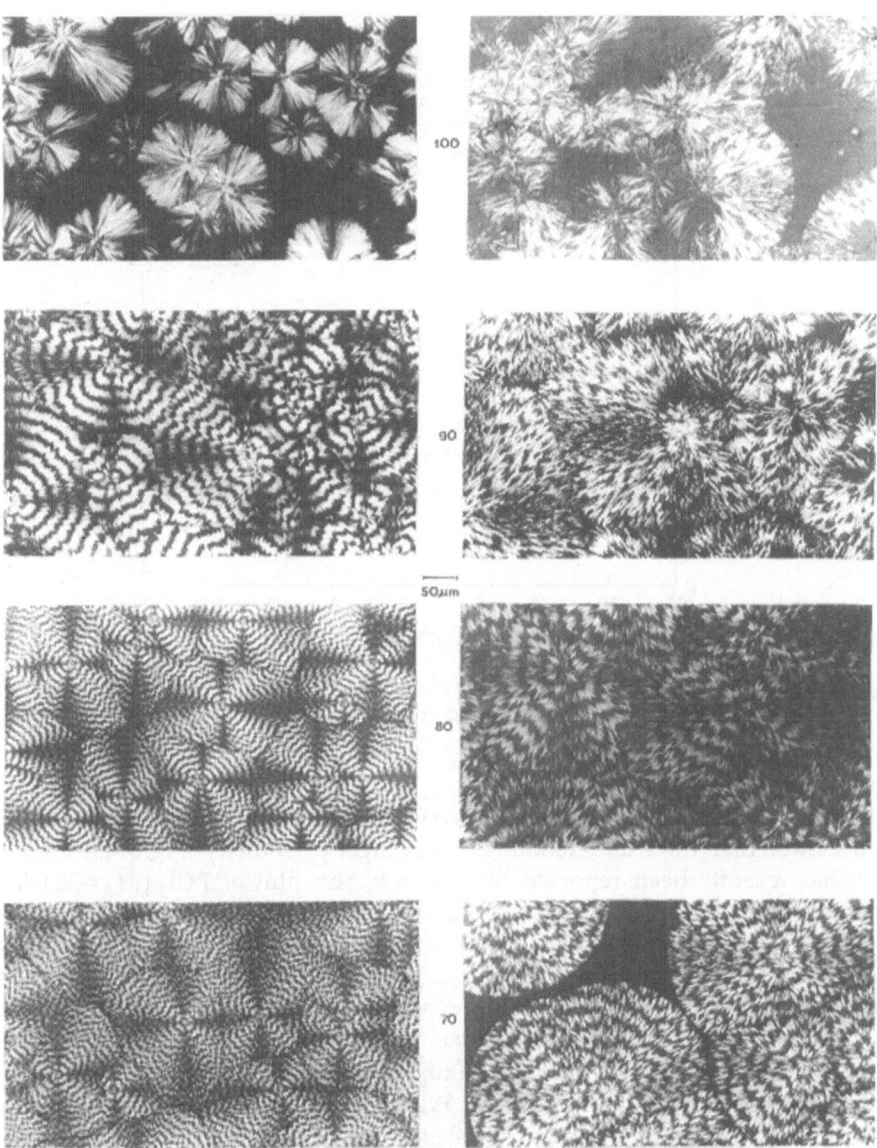

Fig. 28. Optical micrographs of PCL and PCL SAN blends crystallised at 45° C (*left column*) and 50° C (*right column*); central numbers indicate PCL contents (wt %); taken from [106]

perature (T_c) and T_g of mixtures. They plotted the variation in periodicity with (T_c–T_g) and claimed that the points lay on a master curve, inferring that periodicity is inherently related to mobilities of segments in mixtures (Fig. 29). Undercooling also affected spherulite structures and they plotted the variation in

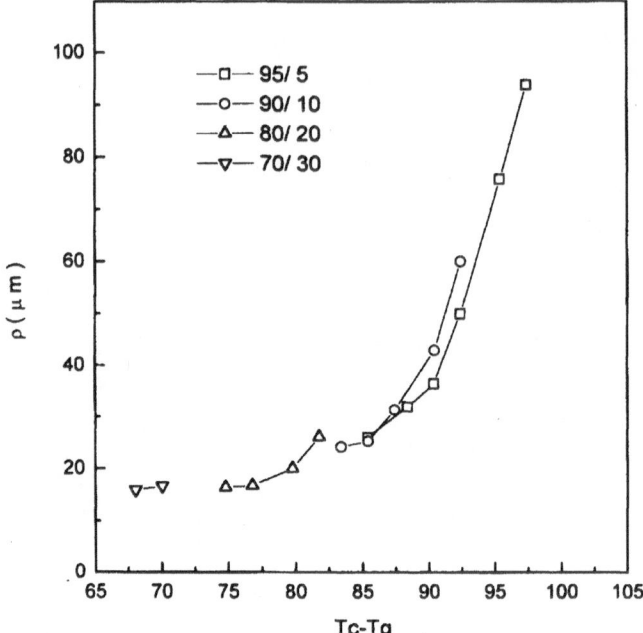

Fig. 29. Variation in periodic distances in ring-banded spherulites in PCL/SAN blends with $(T_c - T_g)$ for different compositions; data taken from [107]

periodicity with $(T_m^0 - T_c)$, where T_m^0 is equilibrium melting point for mixtures, and showed that lower undercoolings gave longer periodicity (Fig. 30).

It has recently been reported that when a thin film of PCL ($\overline{M}_w$=22,000) blended with SAN-25 ($\overline{M}_w$=197,000) was cast from solution in dichloromethane onto glass plates, and examined by scanning electron microscopy, spherulites with two different surface textures were observed. Both sets of spherulites appeared to show ring-banding but one set was relatively light and the other dark in appearance. Wide-angle X-ray diffraction and DSC failed to show the presence of two types of crystalline phase and both sets of data were indistinguishable from those for PCL homopolymer. With some evidence from transmission electron microscopy of replicated surfaces, Wang et al. concluded that the original observations were a consequence of one set of spherulites being nucleated at the glass-polymer interface and the other at the polymer-air interface; the two growth directions giving different textures at the top surface examined [108].

Rates of radial growth of spherulites, constant with time, were highest for pure PCL and decreased with increasing SAN content and with increasing temperature in the range 34–50 °C [106]; i.e. the presence of SAN decreased the rate of crystallisation. The variation in extent of crystallinity with time was sigmoidal and the kinetics of crystallisation were consistent with the Avrami equation (Eq. 26) with an exponent of 3±0.02, consistent with three-dimensional growth

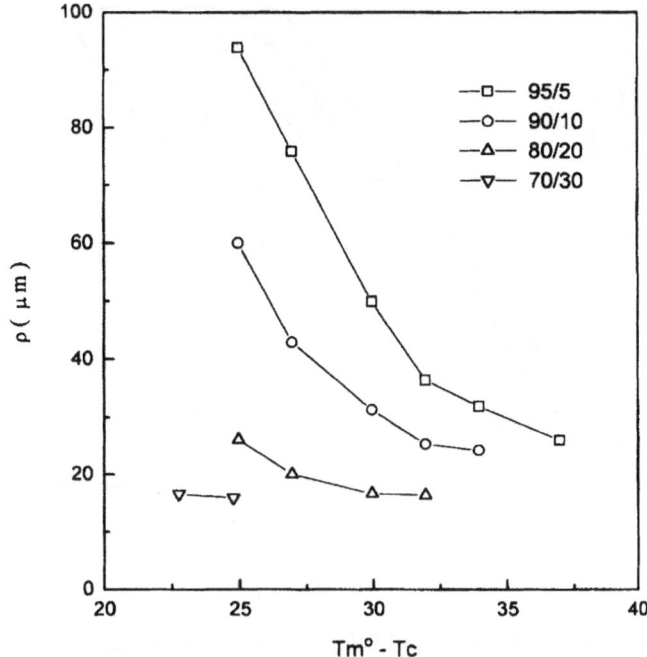

Fig. 30. Variation in periodic distances in ring-banded spherulites in PCL/SAN blends with $(T_m^0 - T_c)$ for different compositions; data taken from [107]

Table 8. Values of the rate coefficient k in the Avrami equation for crystallisation in blends of PCL with styrene-acrylonitrile copolymers (SAN) at 45 °C; data taken from [106]

Wt % SAN	0	10	20	30	40
$10^3 k/min^{-3}$	11.5	4.44	1.15	0.11	0.016

and zero-order nucleation (Fig. 31) Values of k in the Avrami equation for crystallisation at 45 °C are given in Table 8. The authors also considered the kinetics of crystallisation in terms of the Hoffman-Lauritzen secondary nucleation theory. It was stated that this latter analysis indicated increasing disorder in the spherulites when SAN is present.

Kressler et al. [106] also determined the melting behaviour of the spherulites. They determined the variation in T_m with T_c and found very large melting point depressions; extrapolation of T_ms of samples crystallised at different T_cs to $T_m = T_c$ led to an equilibrium melting temperature of about 85 °C which is far greater than the usually accepted value of 71.5 °C; observed melting points were non-linear with T_c and varied with time of crystallisation, indicating some complexity in the system.

Li et al. [80] investigated the kinetics of crystallisation of PCL in PCL/SAN-24 blends (PCL $\overline{M}_w = 101,000$, $\overline{M}_n = 50,600$; SAN-24 $\overline{M}_w = 487,000$, $\overline{M}_n = 145,000$).

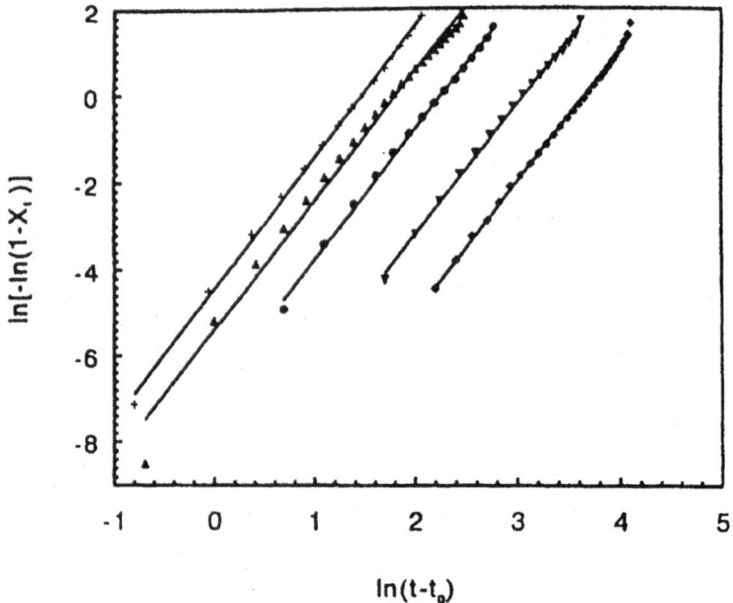

Fig. 31. Linearised Avrami plots for PCL/SAN blends crystallised at 45° C; compositions: +100%, ▲ 90%, ● 180%, ▼ 70% and ◆ 60% PCL; taken from [106]

They determined rates of crystallisation at several temperatures as a function of blend composition. Samples were cast from dichloromethane, heated under vacuum at 40 °C for 48 h. Crystallisation kinetics for blends containing more than 60 wt % PCL were determined after melting, heating to 80 °C for 10 min and rapid cooling to the crystallisation temperature (37.0–45.0 °C). Temperature dependencies of spherulite growth rates were similar for pure PCL and PCL/SAN blends. The spherulites formed had constant rates of radial growth which decreased with increasing SAN content. These data were consistent with, in individual samples, a constant composition at the growth tip of the fibrils throughout the crystallisation process and incorporation of SAN into the amorphous portion of the spherulites.

Crystal melting temperatures were also observed to increase with increasing T_c for a given blend composition. Hoffman-Weeks type extrapolations gave values of equilibrium melting points (T_m^0) for each composition which decreased with increasing SAN content (Fig. 32).

From the several data obtained, Li et al. estimated the PCL/SAN-Flory interaction parameter (χ), the PCL crystal surface free energy ($\sigma_e/(\text{erg cm}^{-2})$) and the product $\sigma\sigma_e$ where σ is the lateral surface free energy [80]. Values of χ were approximately independent of SAN content up to 30 wt % with $\chi=-0.33$. Values of σ_e and $\sigma\sigma_e$, given in Table 9, decrease with increasing SAN content. The reductions in growth rate were consistent with a combination of reduced PCL concentration at the growth surface, an increase in viscosity of the amorphous material due to the

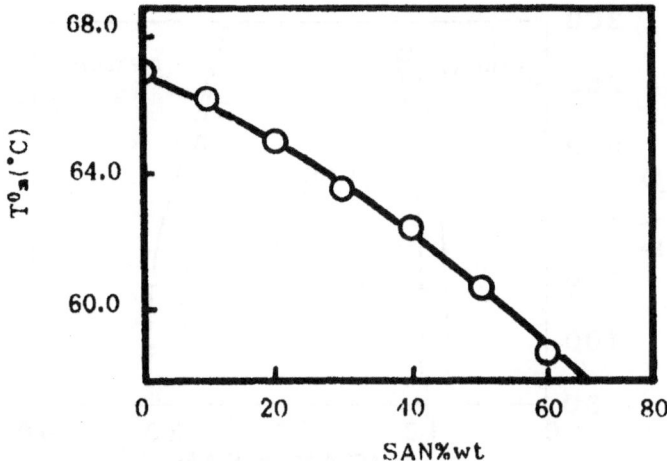

Fig. 32. Variation of equilibrium melting temperature of PCL in PCL/SAN blends as a function of SAN content; taken from [80]

Table 9. Values of PCL crystal surface free energy ($\sigma_e/(J\ cm^{-2})$) and its product ($\sigma\sigma_e$) with σ, the lateral surface free energy in blends of PCL with SAN; data taken from [80]

Wt % SAN	0	10	20	30
$10^7\sigma_e/(J\ cm^{-2})$	85.7	83.6	77.9	71.9
$10^9\sigma\sigma_e/(J^2\ cm^{-4})$	7.54	5.60	5.22	4.82

Table 10. Molecular weight and composition data for SAN copolymers used in PCL blends; data taken from [79]

Composition	$\overline{M}_w$	$\overline{M}_w/\overline{M}_n$
SAN-12.4	142,000	2.11
SAN-14.9	141,000	2.60
SAN-19.5	130,000	2.50
SAN-21.9	165,000	2.08
SAN-26.4	168,000	2.21

presence of SAN (with a higher glass-transition temperature) and also the favourable energy of interaction for mixing between the components, which reduced the free energy change on crystallisation and driving force for crystallisation.

Kressler et al. [79] investigated the kinetics of PCL-700 crystallisation with a series of SAN samples with different AN contents; see Table 10. Blends were prepared by casting from dichloromethane and drying under vacuum at 60 °C. The phase diagram (Fig. 33) shows the combined results of Chiu and Smith [20] and Shulze et al. [109]. For blends with different compositions, constant rates of ra-

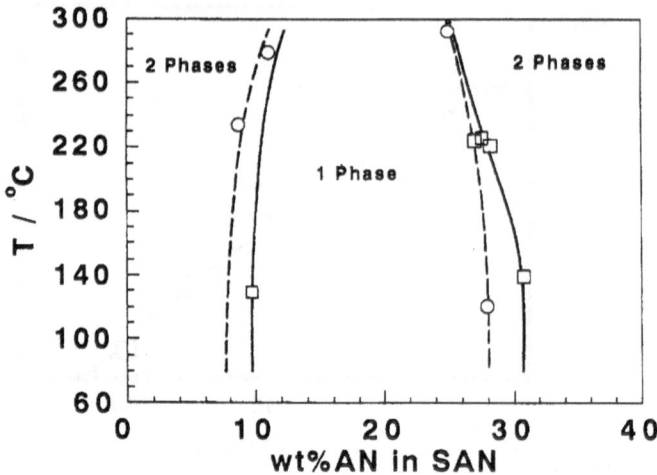

Fig. 33. Phase diagram showing miscibility window in PCL/SAN blends (30 wt % SAN) with different SAN compositions; data taken from [20] (O) and [109] (□); taken from [78]

dial growth of spherulites were observed; rates decreased as SAN content increased (as in other studies described above) with other conditions held constant. As in previous studies the rates decreased with increasing temperature from 35 °C to 50 °C, the only realistically accessible range; below 40 °C growth was too fast and above 50 °C the crystallisation was too slow for convenient measurements. Kressler et al. [79] also analysed the data in terms of Hoffman-Lauritzen secondary nucleation theory; an additional term relating to molecular mobility was introduced. On the basis of that analysis, they estimated values of surface free energy $(\sigma\sigma_e)^{1/2}$ J cm^{-2} as a function of blend composition at a series of SAN compositions. The surface free energy of the crystallites decreased as SAN content increased (Table 11) but was little affected by SAN composition except at high (40 wt % SAN) for SAN-12.4, where a distinctly lower value was obtained; the results imply that the crystal surface in the lamellae became more stable in the presence of SAN. They estimated values of the interaction parameter χ for the several systems and showed that both rates of radial growth of the spherulites are a minimum and χ is low at about 20 wt % AN in SAN (Table 5). They interpreted these observations as indicating that strong polymer-polymer interactions reduce the rate of radial growth by retarding the reptation of polymer chains towards the growth surface of the spherulites. During the course of this study they observed that equilibrium melting points varied with times of crystallisation and Hoffman-Weeks plots, used to determine equilibrium melting points, were not satisfactory. Also, from secondary nucleation analysis, they concluded that there was some evidence for the existence of UCST behaviour in the system at temperatures below the crystal melting temperature of PCL.

More recently, Wang and Jiang [110] investigated the kinetics of PCL ($\overline{M}_w$= 22,000, $\overline{M}_n$=11,300) crystallisation in blends with SAN-25 ($\overline{M}_w$=197,000, $\overline{M}_n$=

Table 11. Surface free energies $10^7\sigma\sigma_e/(\text{J cm}^{-2})$ as functions of SAN composition and content in PCL blends; data taken from [79]

SAN content (wt %)	SAN-12.4	SAN-14.9	SAN-19.5	SAN-21.9	SAN-26.4
10	24.60	23.99	23.82	23.97	24.37
20	23.47	23.21	22.51	23.19	23.00
30	22.26[a]	22.04	22.77	21.87	22.57[a]
40	17.68[a]	21.39	22.17	21.28	20.92[a]

[a] Values at intermediate rates of super-cooling

106,000) cast from dichloromethane and dried under vacuum at 40 °C for 3 days. These workers observed general features of the crystallisation kinetics in agreement with those of other workers. They analysed their data in terms of a modified Turnbull-Fisher theory which predicts the usual bell-shaped curve for crystallisation kinetics in polymers in which growth rate G_m is nucleation controlled at low undercooling and diffusion controlled at high undercooling:

$$G_m = \varphi_2 G_0 e^{-\Delta E/R(T_c-T_g+C)} e^{-\Delta F_m^*/k_B T_c}$$

(27)

where

$$\Delta F_m^* = \frac{2b\sigma\sigma_e}{\Delta h_u f\left(1-\dfrac{T_c}{T_m^0}-\dfrac{RT\chi}{\Delta h_u f}\dfrac{V_{2u}}{V_{1u}}\left(1-\varphi_2\right)^2\right)}$$

and ϕ_i, V_{ui} are the volume fractions and molar volumes of the i-th component, G_0 is a constant dependent on the crystallisation regime, ΔE is the energy to transport segments across the solid-liquid interface, T_c, T_g and T^0_m are the temperatures of crystallisation, glass-transition of the mixture and the equilibrium melting point for the pure crystallisable component, C was taken to be 105 °C (to give best fit with data), ΔF^*_m was derived previously from a lattice treatment, b is the thickness of the critical nucleus, $\sigma\sigma_e$ is the product of the product of the free energies of the lateral and fold surfaces, Δh_u is the heat of fusion per mole of units of the crystallisable component, χ is the Flory-Huggins interaction parameter and f is given by $2T_c(T_c+T_m^0)$. They adopted a value for T_m^0 of 67.1 °C, determined from Hoffman-Weeks plots for their PCL sample and determined a value for G_0 of 134.7×10^{-8} µm min^{-1} by curve fitting. Calculated growth rates were compared with experiment and good agreement was obtained (Fig. 34), except for some deviation at the largest degrees of undercooling (below 30 °C for the blend with 90% PCL). It was found that the agreement was insensitive to differences in parameters, using values of –0.34 and –0.18 reported by others. Values of $\sigma\sigma_e$ were very similar for the two values of χ used (675.3×10^{-6} and 675.1×10^{-6} J^2 m^{-4}, respectively for blends with 10% SAN-25 and 631.0×10^{-6} and

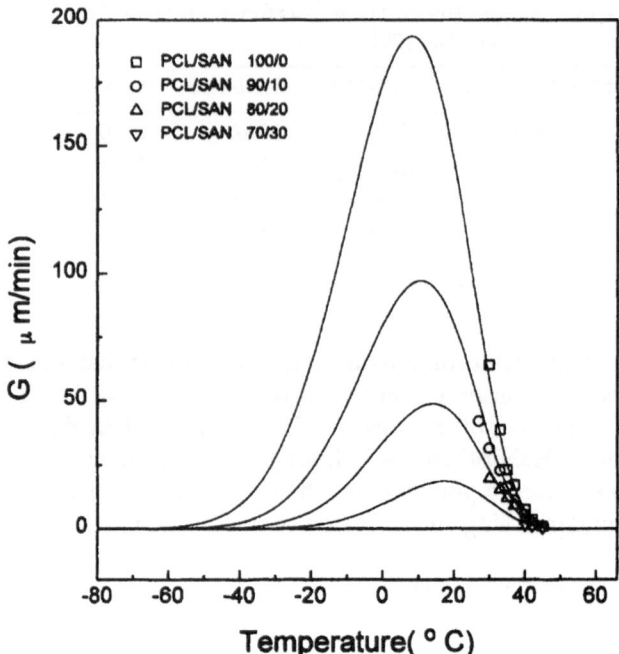

Fig. 34. Comparison of experimental and calculated spherulite growth rates in PCL/SAN blends; curves calculated using a value of –0.34 for the interaction parameter; taken from [110]

$629.1 \times 10^{-6} \, J^2 \, m^{-4}$ for blends with 30% SAN-25; cf. Table 9) and it was suggested that the free energy of critical nucleus formation was not effectively modified by the presence of the amorphous component. Because values of T_g affect kinetics they used slightly modified T_gs and found that calculated growth rates were more sensitive to this parameter; they concluded that T_g is the major factor controlling growth rates of spherulites [110].

Kummerlöwe and Kammer recently further reported on the nature of ring-banded spherulites in PCL/SAN blends. Having suggested that the periodicity L in these structures is given by Eq. (19), they reported data in support of this relationship. For blends of PCL with SAN copolymers of different composition and of different proportions of PCL and SAN, they plotted data which are in accord with Eq. (19), Fig. 35 [52].

While several groups have investigated crystallisation from homogeneous amorphous melts it is also feasible to explore crystallisation from phase-separated liquids.

In the central region of the miscibility window blends are miscible up to the thermal decomposition temperature of PCL and phase separation on heating is not observed. However, at the edges of the miscibility window the shape of the window is such that blends which are miscible at low temperatures show LCST

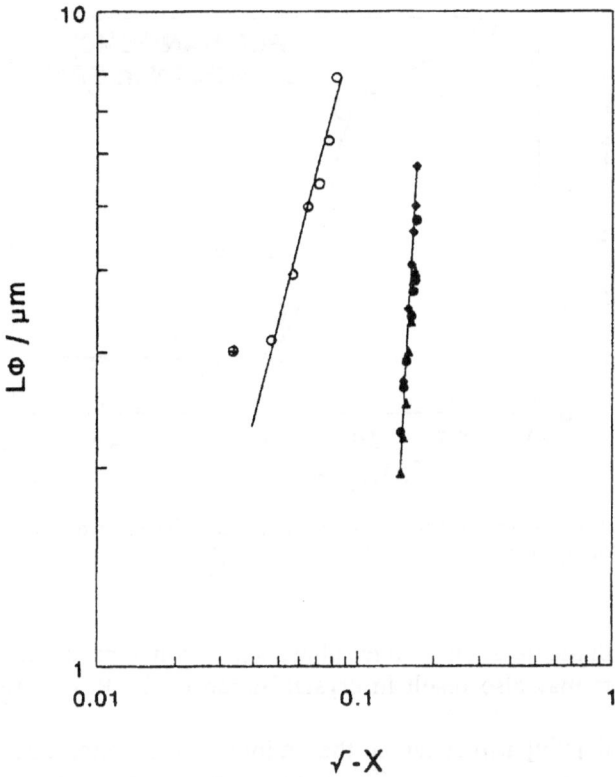

Fig. 35. Variation in the product of the periodicity L and volume fraction ϕ of SAN as a function of the square root of $(-\chi)^{1/2}$, where χ is the interaction parameter and is negative, in PCL/SAN blends; PCL/SAN-12.9 ○ 50:50; PCL/SAN-22.7 ◆ 70/30; ▲ 60/40; ● 50:50; taken from [52]

behaviour on heating and segregate by liquid-liquid phase separation. Schulze et al. [109] investigated the crystallisation of such blends which were cooled to room temperature after allowing liquid-liquid phase separation to proceed for various times. Under these circumstances liquid-liquid phase separation proceeds by spinodal decomposition in which concentration fluctuations develop with a characteristic length scale d; the fluctuations create discontinuous PCL-rich and SAN-rich regions and the difference in compositions of these regions increases until segregation into phases of equilibrium composition is complete; as phase separation continues the phase structure may coarsen (d increases and the discontinuous structure may disintegrate to form droplets of the minor phase).

If PCL then crystallises, on cooling, in the PCL-rich regions, the "spherulites" grow to a diameter D (D<d). Schulze et al. [109] observed such behaviour by optical microscopy and found numerous small crystalline nuclei separated by SAN-rich regions after crystallisation at room temperature. Detailed morpholo-

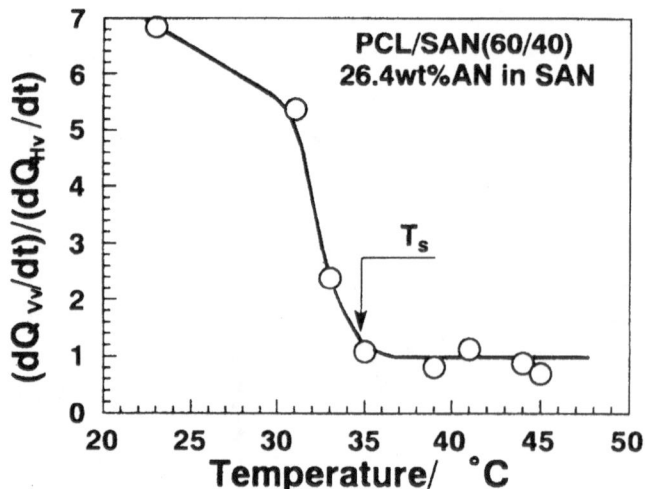

Fig. 36. Temperature dependence of a ratio of parameters derived from scattering data, as described in the text, for PCL/SAN blends; taken from [111]

gies varied with the times allowed for phase separation to proceed. Certain thermal treatments may also result in crystallisation of the PCL in the SAN-rich phase.

Schulze et al. [109] also observed that in immiscible blends, after crystallisation at higher temperatures (40 °C and above) where large spherulites formed, SAN inclusions were contained in spherical domains interspersed within the PCL spherulites. In miscible blends, where SAN was incorporated in the interlamellar regions of the spherulites, ring-banded spherulites were formed. At high crystallisation temperatures (50 °C) the ring-banded structure, arising from (co-operative) twisting of the lamellae, was irregular while at lower temperatures (45 °C) the structure was more regular. Ring periodicity increased with PCL content, rising rapidly at very high PCL contents (extrapolating to greater than the spherulite size in pure PCL where ring-banding was not observed, a result similar to that observed with PVC blends as discussed above) and was more pronounced at high rates of crystallisation; slow rates probably allowed relaxation of the forces responsible for the lamellar twisting responsible for the effect, but detailed mechanisms remain uncertain.

While the existence of LCST behaviour in PCL/SAN blends at the edges of the miscibility window is well established, some evidence became available (from studies of crystallisation kinetics [79] described above) that a UCST might also exist in the same system, but at lower temperatures. Svoboda et al. [111] investigated the crystallisation of PCL in blends with SAN after rapid quenching of thin (20 mm) films from the homogeneous melt. (PCL: $\overline{M}_w$=40,400, $\overline{M}_w/\overline{M}_n$=2.61, SAN with different S/AN ratios: $\overline{M}_w$~140,000, $\overline{M}_w/\overline{M}_n$~2.2). These workers used different light scattering techniques to monitor the physical processes oc-

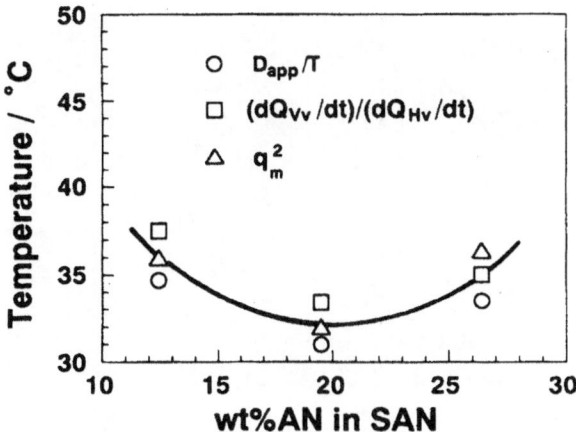

Fig. 37. Phase boundary, showing LCST behaviour, in PCL/SAN blends with 60 wt % PCL as a function of SAN composition; taken from [111]

curring. The techniques involved monitoring scattered light using different optical geometries aimed at distinguishing between the development of anisotropic fluctuations and of density and orientation fluctuations.

A ratio of parameters determined from the scattering data was calculated and shown to vary very markedly with changes in the temperature of crystallisation [111]. This parameter, which was a ratio of rates of change of scattering intensities for two optical configurations and was a measure of relative rates of crystallisation and phase separation, had a value of unity (Fig. 36) at values of T_c above 35 °C for PCL/SAN-26.4 ($\overline{M}_w$=168,000, $\overline{M}_w/\overline{M}_n$=2.21), 60 wt % PCL, as it did for the crystallisation of pure PCL. Below 35 °C the ratio increased markedly to values of 6 or 7. These data were interpreted to imply that, at low temperatures, liquid-liquid phase separation preceded PCL crystallisation and that the temperature at which the ratio of parameters started to increase was the spinodal decomposition temperature (T_s) for that separation. That is, the authors claim that above T_s slow crystallisation of PCL proceeded from the homogeneous melt but that below T_s liquid-liquid phase separation commenced by spinodal composition with the development of concentration fluctuations and that crystallisation of PCL proceeded from the PCL-rich regions which developed. Three different methods of estimating T_s gave consistent results for three different SAN compositions; PCL/SAN 60/40 in all cases (Fig. 37). That is, these data support the concept of liquid-liquid phase separation and, hence, the existence of a UCST below the region of miscibility but in a region not normally accessible experimentally.

The analysis of the scattering data was supported by transmission electron microscopy of stained (RuO$_4$) ultrathin sections. Blends crystallised at the higher temperature showed the existence of extended thin PCL lamellae interspersed by a SAN-rich phase. A sample crystallised at low temperature, below the UCST, showed PCL to be in small irregularly-shaped regions of roughly uniform di-

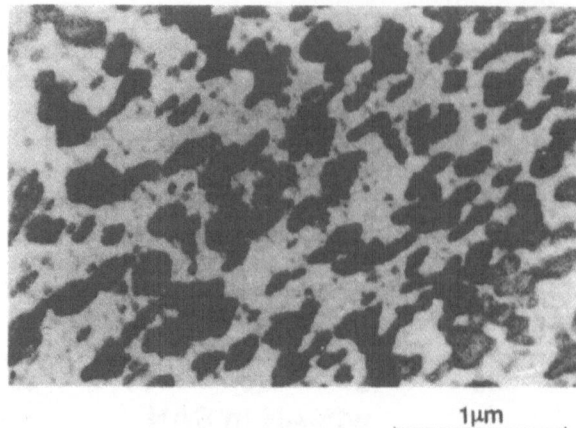

Fig. 38. Transmission electron micrograph of a sample of a PCL/SAN-12.4 blend (60 wt %
PCL) quenched at 24° C; taken from [111]

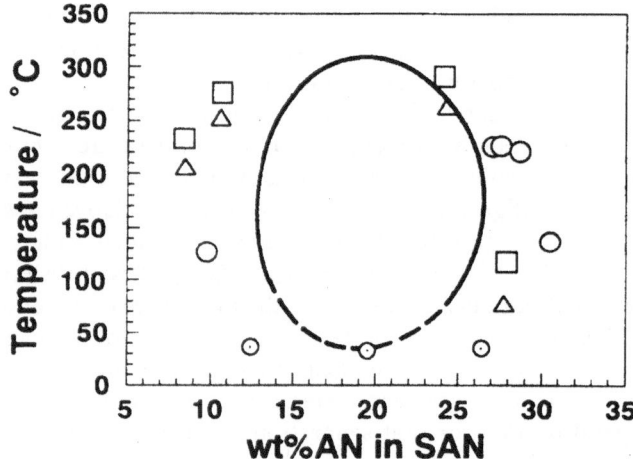

Fig. 39. Miscibility map for PCL/SAN blends showing the variation in phase boundary with
SAN composition: (⊙) UCST data reported in [90], (□) data from [78] (30 wt % PCL),
(△)data from [78] (50 wt % PCL), (○)data from [89] (30 wt % PCL); taken from [111]

mensions (consistent with spinodal decomposition); the internal texture of the
crystalline PCL regions could not be determined (Fig. 38) [111]. It is uncertain
as to what extent low-temperature segregation of the components had proceeded prior to PCL crystallisation.

Svoboda et al. [111] used an extended form of the Flory-Huggins equation for
the free-energy of mixing, in which the interaction parameter was replaced by a
generalised free-energy parameter, to calculate a miscibility map for PCL/SAN

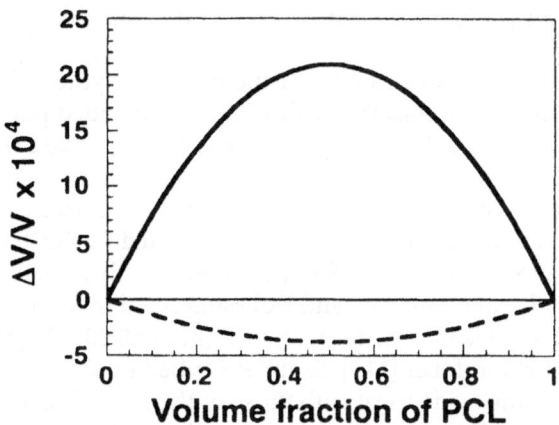

Fig. 40. Volume changes on mixing for PCL/SAN-22 at 85° C (*solid line*) and 260° C (*broken line*); taken from [111]

blends; values of the several segment-segment interaction parameters were estimated and used in the calculations. The resulting miscibility map in terms of temperature and AN content in SAN showed a closed loop (Fig. 39). The calculated loop is in general agreement with previously determined LCST data [20] and the newer UCST data [109].

Svoboda et al. [111] referred to the UCST as a virtual UCST. UCST behaviour was only observed in rapidly quenched samples where samples were cooled so quickly that they were maintained as mobile liquids below the normal phase separation/crystallisation temperature. In bulk samples, where cooling was restricted by the thermal conductivity of the sample, it is unlikely that such phenomena will be observed.

The same authors also estimated volumes of mixing (Fig. 40) and calculated that at high temperatures (260 °C) the volume change on mixing is negative but that at lower temperatures (85 °C) the volume change is positive and suggested that these negative and positive volume changes are connected with LCST and UCST behaviour, respectively.

12
Blends with Aliphatic Polyesters

Poly(ε-caprolactone) is one member of a substantial series of aliphatic polyesters with structure 2 with different values of n. It is therefore pertinent to examine the behaviour of blends of PCL with other members of that family. These systems will have the same types of intermolecular interactions, i.e. methylene-methylene, ester-ester and methylene-ester, but in different proportions. There is also considerable potential interest in their blends because such polymers are biodegradable. However, results show that, in general, the polymers are mutually immiscible and there have been several studies on the behaviours of the blends.

Rocha et al. [112] demonstrated that the parent polymer of the series, poly(glycolic acid) n=1, is immiscible with PCL. They blended poly(glycolic acid) with PCL-767 and PCL-787 (Table 1) by freeze drying mixed solutions under vacuum, mixing the polymers in a Haake rheometer at 235 °C and then compression moulding. Two T_gs were observed by DMA, unshifted from those of the homopolymers; drastic phase separation was reported for 50:50 blends.

Poly(3-hydroxybutyrate) (PHB) 8 was a commercially available, biodegradable, non-linear polyester. Kumagai and Doi established that this polymer ($\overline{M}_w$= 652,000, $\overline{M}_w/\overline{M}_n$=1.8) is immiscible with PCL ($\overline{M}_w$=68,000, $\overline{M}_w/\overline{M}_n$=1.9) when solvent cast from chloroform [113]. Samples studied by DSC showed two glass-transition temperatures, identical with those of the individual components and invariant with composition. Mechanical properties of the blends were poor and tensile modulus and strength were minimal at 50 wt % of the components. PCL had a complex and accelerating influence on the rate of enzymatic degradation of PHB, the kinetics of which were correlated with scanning electron microscopy observations.

PHB is an expensive polyester which crystallises slowly from the melt and embrittles on ageing, while PCL is said to have a rather low melting point and limited use for some applications. In an attempt to produce blends with a more optimal combination of cost and properties, Gassner and Owen [114] examined blends of PHB ($\overline{M}_n$=222,000) with PCL (TONE P-767, Table 1). The polymers were mixed and compression moulded at 190 °C, quenched from the melt and then allowed to crystallise at room temperature for some weeks. They found that the polymers were mutually immiscible and underwent phase separation in the melt to produce phase-separated, partially crystalline materials; each polymer crystallised independently. Crystal melting endotherms (PCL 63 °C and PHB 180 °C), observed by differential thermal analysis (temperature rise 5 °C min^{-1}), were found to grow or shrink with composition. These workers used literature data for heats of fusion of the two polymers and estimated that the crystallinities of PCL and PHB were 57% and 75%, respectively. Slight shifts in melting points were observed with changes in composition but no glass-transition temperatures were observed.

Dynamic mechanical properties were also determined when, for samples containing 60 wt % or more PCL, glass-transition temperatures for PCL blends were observed and samples flowed above 60 °C [114]. Samples with higher PHB

Structure 8

contents retained their properties up to PHB melting. The results suggested that, for samples containing more than 60 wt % PCL, the PCL formed a continuous semicrystalline matrix with embedded PHB spherulites. For samples with less than 40 wt % PHB the PHB became continuous and the PCL did not drastically reduce sample rigidity. Tensile experiments showed that samples with a PCL matrix were ductile while those with a PHB matrix were brittle. For samples high in PCL the PHB increased the overall modulus due to the reinforcing effect of high-modulus inclusions.

Scanning electron microscopy showed that samples with a PCL matrix developed longitudinal cavities when drawn to 20% as result of PHB inclusions not deforming. Samples containing more than 60 wt % PHB underwent brittle fracture and studies of freeze-fractured samples showed a layered structure parallel to the sample surface.

Poly(3-hydroxybutyrate) is also available modified with hydroxyvalerate (HV) as copolymers with different amounts of HV units. Yasin and Tighe investigated blends of PCL ($\overline{M}_w$=50,000) with Biopol PHB, modified with 12% ($\overline{M}_w$= 350,000) and 20% ($\overline{M}_w$=300,000) HV; both of these polymers are crystallisable. Samples were solvent cast from methylene chloride or methylene chloride/THF mixture or by injection moulding at about 150 °C [115]. Solvent-cast blends containing <10% PCL, studied by scanning electron microscopy, were reported to be compatible, although even at low PCL contents small PCL spherulites (which increased in size with increasing PCL content) could be found by optical microscopy. Opacity increased with PCL content, possibly due to PCL crystallisation. Quenching of samples heated to 120–130 °C gave rise to clear films but on slow cooling the PCL crystallised. Gross phase separation was reported for samples with 50% of each component [115].

PCL was also blended with poly(3-hydroxybutyrate-co-3-hydroxyvalerate) (PHBV) copolymers on a small two-roll mill for 5 min at 140 °C (12% HV) or 128 °C (20% HV). The blends were injection moulded to produce tensile test samples. Samples were prepared in the presence of 1 wt % hydroxyapatite as a crystal nucleating agent and to reduce sample preparation time. For both copolymers, the addition of PCL (10 wt %) reduced the initial modulus and tensile strength (Table 12). Samples containing 90 wt % of PCL exhibited properties very similar to those of PCL and the effects were attributed to phase separation from the blends brought about by PHBV crystallisation at temperatures exceeding T_m for PCL. That is, the samples show immiscibility. Even blends containing 10 wt % PCL samples annealed in aqueous media or in an oven showed the presence of crystalline PCL.

Samples prepared as above were subjected to accelerated hydrolytic degradation at 50 °C. Interpretation of the results was complicated by the different crystalline textures and large differences in molecular weights of the components. Nevertheless, the polymers degraded most rapidly under alkaline conditions. The deterioration in tensile properties with degradation at pH 7.4 and 37 °C was monitored [115].

Block or graft copolymers (which have the same influence as block copolymers as interfacial agents) may also be prepared in situ, not by transesterifica-

Table 12. Tensile properties of blends of PCL with PHB-PHV copolymers containing 12% and 20% PHV; data taken from [115]

Copolymer	Wt % PCL	Initial Modulus (MPa)	Extension at yield (%)	Yield stress (MPa)	Ultimate stress (MPa)	Elongation at break (%)
–	100	150±20	18±6	16.3±1.2	20.1±1.2	400±60
12% PHV	90	150±10	43±5	15.0±0.8	15.0±0.8	43±5
20% PHV	90	150±10	31±2	15.3±0.3	15.3±0.3	31±2
12% PHV	10	300±30	13±2	19.9±0.1	19.90.1	13±2
20% PHV	10	300±40	18±4	18.0±1.2	19.20.7	400±50
12% PHV	0	500±30	12±2	28.7±1.8	27.11.8	15±2
20% PHV	0	310±10	13±2	19.1±0.2	17.30.9	22±2

tion but by reactive blending. An example of such a system involving PCL has been given by Cavallaro et al. [7, 8]. PHBV and PCL were blended in the presence of a peroxide. Decomposition of the peroxide generated radicals which could abstract a hydrogen atom from a polymer, probably the PHBV, to generate a macroradical and then, through transfer reactions, PCL macroradicals could be formed and radical-radical recombination reactions of macroradicals could generate graft copolymers.

PCL (M_v=50,000) was blended with PHBV, containing 4% HV, in the proportions 70:30 and 30:70 (w/w). Both polymers alone are crystalline. Blends were prepared by simple melt mixing and by reactive blending, i.e. by melt mixing in the presence of a peroxide which generated free radicals during the mixing process, at 100 °C and at 160 °C [8]. Heating the individual polymers in the presence of peroxide at 160 °C reduced the melting point of PCL, from 64.3 °C to 60.4 °C, but raised the melting point of PHBV from 159.2 °C to 172.1 °C[8].

In the blends melting points for PCL were depressed to about 60 °C and did not change with reactive blending at 100 °C or at 160 °C. For samples prepared by reactive blending at 160 °C the melting point for PHBV was raised by about 10 °C in the blend containing 70% PHBV but was unaffected in that containing 30% PHBV. In addition, when samples containing PHBV were heated in the presence of peroxide at 160 °C they became insoluble in chlorinated solvents while PCL was unaffected. The results were consistent with the crosslinked samples having a PHBV matrix.

In general, the extent of crystallinity of each component decreased as its content in the sample decreased, except for PCL in blends prepared by reactive blending at 160 °C, when the crystalline content increased slightly as its overall content decreased. It was suggested that these results might be explained by miscibility of the polymers in the melt and rejection of PCL in the liquid state as PHBV crystallised in blends where PHBV is not crosslinked but restricting segregation of PCL in cases where the PHBV is crosslinked and producing more extensive crystallisation of PCL in the interstices of the PHBV crystallites.

Table 13. Tensile properties of blends of PCL with PHBV with and without reactive blending at high and low temperatures; data taken from [8]

Wt % PCL	Blending conditions Temperature (°C)	Reactive	Initial modulus (kg cm^{-2})	Ultimate tensile stress (kg cm^{-2})
100		no	3.6	5.01
100		yes	2.3	4.20
73	100	no	4.6	1.70
73	100	yes	5.3	1.29
73	160	no	4.2	1.30
73	160	yes	5.2	1.83
37	100	no	14.8	2.09
37	100	yes	12.7	1.74
37	160	no	15.7	2.11
37	160	yes	11.3	1.43
0		no	20.0	2.22
0		yes	16.0	3.28

Further, differential scanning calorimetric data indicated that, on quenching PHBV from the melt, crystallisation was at best partial; a glass-transition temperature and crystallisation peaks were observed on slow heating, but after crosslinking the crystallisation on cooling was far stronger and only a crystal melting peak was observed on subsequent heating. All samples exhibited two detectable T$_g$s, one for each component, which showed no trend with sample type [8].

Tensile moduli of the blends were intermediate between those of the components and were dominated by the properties of the matrix component (Table 13). The moduli were reduced by reactive blending compared with simple melt mixing. Tensile strengths of blends were significantly reduced, and were especially reduced for those with PCL as the major component; the tensile strength of PHBV is lower than that of PCL [8].

Fracture surfaces of blends with PHBV as the major component showed the presence of PCL which, in melt mixed samples, had no interaction with the matrix but in reactive blended samples there was plastic deformation at the polymer-polymer interface. Blends with a PCL matrix also provided evidence for interface attachment in reactive blends [8]. Spectroscopic, differential scanning calorimetry and thermogravimetric data were quoted to support the formation of copolymer in this system [7].

A further polyester blend system which has been investigated recently is the combination of PCL with poly(L-lactide). L-Lactide has the structure **9a** which on polymerisation incorporates two ester units per repeat unit into the polymer chain **9b**.

Structure 9a

Structure 9b

in which the main-chain carbons carrying the methyl groups are chiral.

In one study, poly(L-lactide) ($\overline{M}_n$=70,000, $\overline{M}_w/\overline{M}_n$=3.2) and PCL ($\overline{M}_n$= 55,000, $\overline{M}_w/\overline{M}_n$=1.2) were blended in a twin-screw extruder at 200 °C in an attempt to prepare block copolymers through transesterification reactions in the presence of various catalysts [116]. Catalysts were added as 20 wt % solution in toluene. Poly(L-lactide) decomposes slightly above its melting point (180 °C) and the processing window is small; processing with limited residence times at 200 °C was considered satisfactory although there was some decomposition [116]. Coextrusion of the polymers (50/50) with a residence time of 5 min without catalyst gave no transesterification and no lactide formation; i.e. residual catalysts left in the polymers on synthesis did not catalyse reactions.

Addition of 2 wt % n-Bu$_3$SnOMe as catalyst caused some transesterification but with simultaneous elimination of 12 wt % of lactide; lower catalyst contents were not effective in promoting transesterification. Other catalysts, Ti(OBu)$_4$, Sn(2-ethylhexanoate)$_2$, Y(2-ethylhexanoate)$_3$ and p-toluene sulfonic acid, were ineffective; the PCL remained unchanged but there was some poly(L-lactide) degradation. An attempt to prepare block copolymers, by increasing the content of reactive hydroxyl endgroups, was also made by polymerising L-lactide monomer in the presence of hydroxyl-terminated PCL ($\overline{M}_n$=1500, $\overline{M}_w/\overline{M}_n$=1.5). The low-molecular-weight PCL in the presence of Ti(OBu)$_4$ did not promote transesterification.

In an alternative to reactive extrusion, a blend was prepared from solution in chloroform, in the presence of n-Bu$_3$SnOMe. Heating this blend to 150 °C for 72 h (to investigate the effect of extended reaction times but at lower temperatures) gave some low molecular weight random copolyester but a lot of lactide. Use of Ti(OBu)$_4$ as catalyst gave no reaction. As a further alternative, polymers were heated together in solution in diphenyl ether at 170 °C for 120 h. The catalysts Sn(2-ethylhexanoate)$_2$ and ZnCl$_2$ promoted the formation of some copolymer with some poly(L-lactide) degradation; Ti(OBu)$_4$ did not promote any reaction [116].

In a subsequent study, poly(L-lactide) and PCL were prepared separately by polymerisation initiated with ethanol at 110 °C in the presence of stannous octoate as catalyst. Low temperatures were used to avoid any possibility of transesterification which could interfere with attempts to form block copolymer. Attempts were made to prepare block copolymers by reacting first-formed polymer, having reactive hydroxyl end groups, with the second monomer [117].

Addition of L-lactide monomer to preformed active PCL chains gave 98% conversion of lactide monomer and formation of block copolymer. That is, the PCL end groups initiated the ring-opening polymerisation of L-lactide. It was also demonstrated that there was no transesterification reaction during the polymerisation; the end groups of the propagating lactide chain did not attack the preformed PCL chain.

When the procedure was reversed, and ε-caprolactone monomer was added to active poly(L-lactide) chains in the presence of stannous octoate, there was 100% conversion of the ε-caprolactone. NMR evidence indicated that the product of this reaction was a random copolymer; LCL, CLC and LLC sequences were identifiable; (L and C are the sequences from L-lactide and ε-caprolactone, respectively). It was also found that lactide hydroxyl end groups (as in 9b) were present at the end of the copolymerisation in higher content than endgroups from ε-caprolactone (as in 10), which were hardly discernible. Thus, the ε-caprolactone hydroxyl groups, formed during the polymerisation of the caprolactone, were capable of reacting with the preformed ester linkages of the polylactide.

It was also shown that lactide ends would not initiate ring-opening polymerisation of ε-caprolactone in the presence of aluminium triiospropoxide; caprolactone ends 10 would open L-lactide rings [117].

Reaction mechanisms for transesterification reactions were discussed and it was concluded that the ε-caprolactone ends 10 are more reactive than the lactide ends 9b, both in terms of electronic and steric effects. The carbonyl group in close proximity to the end of the lactide chain was considered to reduce the nucleophilicity of the terminal hydroxyl group and the alpha-methyl group was also considered to provide steric hindrance [117].

In a separate study, PCL (molecular weight 15,000) was blended with poly(L-lactide) (molecular weight 10,000) by casting from solutions in chloroform and drying under vacuum at 60 °C for 24 h. DSC scans of samples heated to 200 °C and quenched showed that the T_g of poly(L-lactide) overlapped with the T_m of PCL, which rendered detailed analysis of the thermograms and assessment of polymer-polymer miscibility difficult. However, the cold-crystallisation exotherm of pure poly(L-lactide), at 131 °C, was shifted in the blends containing more than 50 wt % poly(L-lactide) to about 95 °C; this shift was independent of the poly(L-lactide) content. In addition, the crystallisation rate for PCL was

$$\overset{O}{\overset{\|}{\underset{}{---}}}CH_2-CH_2-CH_2-CH_2-CH_2-OH$$

Structure 10

greatly enhanced by the addition of poly(L-lactide). Such evidence was considered in trying to establish the phase behaviour of the blend. It was considered that if the components were miscible the segmental mobility of the PCL could promote crystallisation of the poly(L-lactide), the rate of which should increase with increasing PCL content in the blend. Alternatively, if the polymers were not miscible, the interface could promote nucleation and enhance crystallisation. Microscopy indicated that, at 200 °C, blends containing 10 wt % and 20 wt % PCL were phase-separated and nucleation of the poly(L-lactide) spherulites did not seem to occur preferentially at the domain interfaces. It was thus concluded that the observed effects were a consequence of partial miscibility of the component polymers [118].

Ketelaars et al. recently examined the densities of amorphous blends (low PCL contents) and found that, while T_g data fitted Eq. (23), the densities of the samples were greater than would correspond to a weighted-average [119]. They also showed that, at higher PCL contents, secondary crystallisation did not lead to an increase in density. These workers concluded that a single excess density parameter, indicative of favourable interactions between the components, is present in the amorphous material throughout the composition range.

13
Blends with Cellulose Esters

Apart from main-chain polyesters, cellulose esters constitute an important series of polymers with ester groups. Cellulose itself is a polymer of D-glucose 11, each repeating unit of which carries three hydroxyl groups, each of which can be esterified. The useful properties of these polymers, but properties which make them difficult to process, include high softening and melting temperatures. Combined with low decomposition temperatures, the above properties give a limited processing window unless the polymers are plasticized or processed from solution. Koleske et al. [120] in an early patent stated that PCL and cellulose esters can provide a series of useful blends. They compounded PCL with, separately, cellulose tridecanoate, cellulose triacetate and cellulose diacetate and with acetylated ethylcellulose. Cellulose triacetate and diacetate each gave slightly hazy materials with up to about 25 wt % added PCL; their moduli and strengths were generally decreased.

Structure 11

13.1
Blends with Cellulose Acetate Butyrate

Subsequently, Hubbell and Cooper [58] studied blends of PCL ($\overline{M}>_n$=13,000, $\overline{M}_w$=24,000) with cellulose acetate butyrate (CAB) ($\overline{M}_n$=77,000). They showed that samples rich in CAB (>50 wt %) were optically clear and exhibited no melting endotherm. Blends rich in PCL exhibited a melting endotherm for crystalline PCL. Glass transition temperature data (Fig. 41) did not fit the Fox equation (Eq. 23) or other equations for homogeneous blends. Transition temperatures for quenched samples increased somewhat with increasing CAB content, up to about 50 wt %, and then remained sensibly constant (at about 243 °C) and comparable with glass-transition temperatures of PCL-rich samples when not quenched. These workers suggested that the data corresponded to a two-phase system in the composition range 15–75 wt % PCL, one phase of which contained about 57 wt % PCL (estimated from the Fox equation, Eq. 23). The same workers found a complex system of loss peaks in dynamic mechanical data with no simple trend in E' or E" with temperature; some tensile data were tabulated. Tensile stress-strain curves showed high initial moduli which decreased with increasing

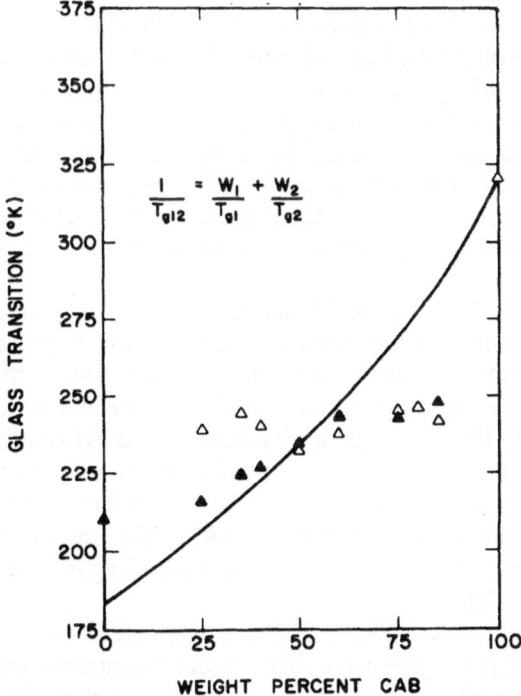

Fig. 41. Variation in glass-transition temperature with composition in PCL/CAB blends for (Δ) unquenched and (▲) quenched samples, with a Fox equation fit for latter data; taken from [58]

PCL content and, above about 20 wt % PCL, a yield which increased with increasing PCL content. The poorest properties were observed for samples with 50 wt % of each component as, under these conditions, there was no PCL crystallinity and PCL reduced the properties of CAB-rich polymers as a result of plasticization. PCL crystallinity seemed to increase slightly on adding up to 35% CAB to PCL and then drop catastrophically to zero at 50% PCL. It was suggested that the presence of incompatible CAB increased the crystallinity in the PCL until the CAB content exceeded about 35 wt % when increasing amounts of PCL were contained within the CAB phase and the CAB prevented it crystallising. Generally the crystal melting temperature of the PCL decreased as the CAB content increased, except that these workers recorded a maximum T_m at about 35 wt % CAB. Small-angle laser light scattering data at low CAB contents indicated space-filling spherulites with PCL chains orientated perpendicular to the radial direction.

Vazquez-Torres and Cruz-Ramos determined glass-transition temperatures by DSC of films cast from methylene chloride; samples were dried for 48 h at 80 °C [121]. Transitions were not clear until samples had been thermally cycled. It was then found that for samples high in CAB (>60 wt %) T_g for CAB decreased with increasing CAB content from 148 °C, for 100% CAB, to 114 °C for 60 wt % PCL. These T_gs were higher than expected for miscible polymer blends when their compositions were corrected for crystalline PCL content. These results disagreed with those of Hubbell and Cooper [58] who observed a T_g of 48 °C for pure CAB and found T_gs of –58 °C to –25 °C for compositions varying from 25 wt % to 80 wt % CAB. Neither group reported two T_gs from a phase separated system and it was suggested that the different results obtained by the two groups arose from different compositions of CAB; the polymer used by Hubbell and Cooper [58] had 1.2 butyl and 1.0 acetyl groups per repeat unit compared with 1.1 butyl and 1.7 acetyl groups in the polymer used by Vazquez-Torres and Cruz-Ramos [121].

For the 50:50 and 25:75 PCL/CAB blends, the latter group found double crystal-melting temperatures which were attributed to two populations of crystals arising from a morphological effect. It was also found that at low CAB (PCL≥50 wt %) in as cast samples the T_m and ΔH_f for PCL were raised relative to those for pure PCL. This effect (which has been found in two other systems) was attributed to an improvement of crystal quality and increased lamellar thickness relative to those found for pure PCL. (This effect is inconsistent with melting point depression but the components are not miscible in the melt.) The effect, which was not explained, persisted after thermal cycling in blends with low CAB contents (20–30 wt %).

Vazquez-Torres and Cruz-Ramos [121] also reported shear modulus and tanδ data for CAB blends; samples were dried under vacuum for one week at 80 °C and were subsequently pressed into samples. T_g for pure CAB was at about 150 °C, in agreement with DSC data. Addition of PCL (25% and 50%) introduced two more relaxations. One was PCL melting at 50–60 °C and a larger peak at about 100 °C attributed to T_g for the blend. The T_g relaxation for CAB persisted

Table 14. Peak temperatures for tanδ data in blends of PCL with CAB; data taken from [121]

Wt % CAB	100	90	75	50	40
$T_1/°C$		130	105	90	80
$T_2/°C$	155	152	152	148	?

but became much weaker; that is there was still a CAB phase, indicating limited miscibility. Above about –50 °C the shear modulus decreased markedly with increasing temperature, to about one third at room temperature and by 1.5 orders of magnitude at >50 °C. Tabulated values of temperatures for the α transition are contained in Table 14.

WAXS data for as-cast blends showed that crystal melting peaks for PCL (relative to PCL contents) were nearly constant; there was a small increase in crystallinity on addition of CAB in line with DSC data. On thermal cycling to 237 °C there was a massive loss of PCL crystallinity as solvent history was lost and the blend took on its own properties, consistent with DSC and differential thermal analysis data. It was noted that in as-cast blends one of the crystal diffraction peaks decreased in relative intensity. This effect was attributed (possibly) to distortion of unit cell dimensions, an effect found elsewhere in paracrystalline materials.

Optical micrographs of PCL in as-cast blends showed well-defined spherulites; there was some distortion of shapes after melting and heating to 237 °C and subsequent cooling. In the presence of 25 wt % CAB the spherulites obtained after heating were more uniform.

In 50:50 mixtures small spherulites were present in as-cast blends but were lost on heating and cooling. Indications were found that there was little crystallinity in 50:50 mixtures or blends with 75 wt % CAB on long annealing at 4 °C, a result said to be consistent with partial miscibility.

PCL/CAB blends, after heating, were examined by NMR and solubility tests. It was concluded that, under the heating cycle employed, no transesterification occurred which might have been responsible for the observed effects [121].

13.2
Blends with Cellulose Diacetate (CDA)

Vazquez-Torres and Cruz-Ramos used solvent-cast samples (cast from methylene chloride/acetone 70/30) dried for 48 h at 80 °C. T_gs in as-cast samples were not well defined but became observable after thermal cycling [121]. The cellulose diacetate T_g decreased slightly, from 180 °C to 176 °C, as the PCL content increased to 60 wt %. The T_g for PCL was not observed and it was concluded that the components were immiscible.

There was some evidence for double melting endotherms for PCL after recycling. T_m for PCL was higher in the presence of CDA (up to 25 wt %) even after thermal cycling. Thus, the presence of CDA has some influence on the quality of

PCL crystallinity but there was a marked decrease in total PCL crystallinity with added CDA in as-cast samples but not much loss in overall crystallinity after thermal cycling.

As-cast CDA blends showed isolated spherulites but these became more uniformly distributed and impinged (with 25 wt % PCL) after heating. CDA appeared to disperse PCL and disrupt and restrict the growth of spherulites; the PCL had to crystallise well below the T_g of CDA and in a rigid matrix.

13.3
Blends with Cellulose Triacetate

Vazquez-Torres and Cruz-Ramos studied blends of PCL with cellulose triacetate (CTA), cast from methylene chloride and dried for 48 h at 80 °C. No glass-transitions were observed and therefore no conclusions on polymer-polymer miscibility were drawn [121]. Double PCL melting peaks were seen and, in this case, the lower T_m was the smaller, cf. CAB blends. Values of T_m and ΔH_f of PCL were not increased by CTA. There was some evidence for CTA crystallinity on heating to high temperatures in the differential scanning calorimeter [121].

In dynamic mechanical analysis experiments T_α for CTA did not shift on addition of PCL but there was marked loss of modulus above –50 °C in the 40:60 and 60:40 blends. Relaxation peaks for T_α and T_m for PCL were also observed.

WAXS data suggested that PCL crystallinity decreased on addition of CTA but increased after heating. Relative magnitudes of various diffraction peaks changed on heating to become closer to the pattern for pure PCL. That is, in as-cast blends the development of PCL crystallinity and morphology was, in some way, restricted.

Optical microscopy showed the presence of small spherulites before and after heating with not much change in their character. There was a significant loss in PCL crystallinity on addition of CTA and these blends were concluded to be immiscible [121].

From the studies of blends of PCL with the several cellulose esters, Vazquez-Torres and Cruz-Ramos [121] suggested that there is some interaction between hydroxyl and carbonyl groups in the different polymers but features did not vary systematically with hydroxyl content and other factors.

14
Blends with Aromatic Polyesters and Copolyesters

A brief report on blends of PCL ($\overline{M}_w$=35,000) with poly(ethylene naphthalate) 12 (molecular weight unspecified), prepared by melt mixing, indicated that the polymers are immiscible; relative crystallinity was a maximum at about 10 wt % PCL. NMR data were obtained for transesterification on heating above 300 °C and transesterification was said to reduce crystallinity in the blends and the value of T_m for poly(ethylene naphthalate) [122].

Structure 12

Structure 13

Poly(ε-caprolactone) ($\overline{M}_w$=40,000) was blended with a series of ethylene terephthalate-PCL copolyesters of different composition 13; the polyesters were presumably random copolymers [123]. The copolymers, containing 18–87% caprolactone units, were prepared by copolymerisation of ethylene terephthalate and ε-caprolactone; molecular weights were not quoted but intrinsic viscosities of the copolymers were between 0.84 dl g^{-1} and 1.99 dl g^{-1}. The pure copolymers exhibited composition-dependent glass-transition temperatures from –65 °C (87 wt % ε-caprolactone) to 33 °C (18 wt % ε-caprolactone).

Blends of PCL with copolymers having high PCL contents (87 wt % and 72 wt %), and quenched from 227 °C, exhibited single glass-transition temperatures which were composition dependent and fitted the Gordon-Taylor equation (Eq. 22) [123]. That is, the PCL-rich copolymers were miscible with PCL at high temperatures. However, when the same samples were cooled at 20 °C min^{-1} from 177 °C the PCL crystallised at temperatures which varied slightly with blend composition, e.g. –13 °C to 7 °C for blends with the copolymer having 87 wt % PCL. Crystal-melting temperatures recorded for PCL melting were almost identical to that of pure PCL except for the blend with low (20 wt %) PCL content, where the crystal-melting temperature was depressed by 10 °C and a low enthalpy of melting was observed. Otherwise enthalpies of crystallisation per gram of PCL increased with increasing copolymer content. The authors, Dezhu et al. [123], attributed the effect to participation of the PCL in the copolymer on PCL crystallisation.

Blends with copolymers having lower PCL contents were less miscible with PCL. Glass-transition temperatures of quenched blends were almost independent of composition and close to those of the copolymers themselves; blend T_gs changed by about 10 °C over the composition range 0–90 wt % PCL. Thus, it is seen that PCL was phase-separated from the copolymers at high temperatures. In such blends the crystal-melting temperatures of samples crystallised as above were very close to those of pure PCL and heats of fusion per gram of PCL

Table 15. Properties of blends of PCL with ε-caprolactone-ethylene terephthalate copolymers; data taken from [123]

Copolyester	PCL content (wt %)	ΔH_m (PCL) $J g^{-1}$	$T_m K^{-1}$	$T_g K^{-1}$
PCL	100	93.18	324	202
TCL-82	0			306
TCL-72	90	85.98	323	280
	80	87.49	324	281
	70	86.27	324	281
	50	81.54	323	284
	20	73.05	320	290
	0			291
TCL-48	90	88.87	323	234
	80	92.34	322	234
	70	89.75	322	235
	50	82.05	321	235
	20	85.96	319	236
	0			246
TCL-28	70			204
	50			210
	30			215
	0			219
TCL-13	90	98.04	323	202
	80	103.73	322	202
	70	100.72	322	202
	50	128.51	320	205
	20	41.65	314	207
	0			208

In TCL-x, x is the wt % ethylene terephthalate content of the ethylene terephthalate/ε-caprolactone copolymers

homopolymer (ΔH_m) were almost constant and equal to that of pure PCL, except at low (<50 wt % PCL) (Table 15) [123].

Optical microscopy of blends containing 80 wt % PCL homopolymer and various copolymers showed that all samples, under a variety of sample histories, were highly crystalline and contained space-filling spherulites. The samples. however, showed considerable differences in detailed morphology which changed dramatically with thermal annealing [123].

Samples containing a copolymer having 87 wt % ε-caprolactone gave small spherulites after crystallisation for 12 h at a variety of crystallisation temperatures; only the samples crystallised at the highest temperature (48 °C) showed any indication of forming ring-banded spherulites [123].

Other copolymers, inherently less miscible with PCL (≤50 wt % ε-caprolactone), showed quite different spherulite features when annealed in the melt at 237 °C for various times and then crystallised at 40 °C for 12 h. A short (5–10 min) heating in the melt increased the sizes of the spherulites formed on crystallisation. The nature of the spherulites also changed and heating times of 80–90 min gave rise to more-regular ring-banded spherulites; repeat distances of the ring-banding decreased with heating times and regularity of structure improved. More prolonged heating (~90 min) brought about a much more diffuse picture of crystalline regions in the melt; well-defined spherulites did not form and details of micrographs varied with the initial copolymer composition [123].

Clearly, heating in the melt brought about changes which strongly influenced subsequent crystallisation behaviour. It was reported that T_gs of samples decreased with increasing times in the melt, up to 40 min. A new glass-transition appeared, attributed to a new component after heating for 60 min. Changes did not occur with heating times longer than 90 min. These changes correlated with changes in the appearance of spherulite structure and were attributed to transesterification in the blends at 237 °C.

Ring-banded spherulites were more commonly seen in blends of PCL with polymers with which there was miscibility in the amorphous phase (cf. PVC and chlorinated polyethylenes, Sects. 8 and 9). The initial copolymers were immiscible with PCL but it was considered that miscibility was enhanced by the formation of different copolymer structures on heating during processing.

A study was also made of blends of PCL with butylene terephthalate-ε-caprolactone copolymers, specifically with a view to investigating transesterification reactions in terms of miscibilities of polymers and crystallisation behaviour [124]. Butylene terephthalate ε-caprolactone copolymers 14, with different caprolactone contents (20–90 wt %) were prepared by reacting with terephthalic acid with two mole equivalents of butane diol and then reacting this product with ε-caprolactone in the presence of a catalyst. These copolymers were blended with PCL ($\overline{M}_w$=40,000) by casting from solution in chloroform and drying under vacuum at 50 °C. DSC was used to investigate samples, using heating and cooling runs at 20 °C min^{-1}, to determine values of glass-transition, crystal-melting and crystallisation temperatures.

In blends of copolymers with 61 wt % caprolactone, one composition-dependent T_g, indicative of miscibility, was found; all copolymers with >40 wt % caprolactone give miscible blends [124]. For PCL blends with copolymers containing 20 wt % ε-caprolactone, DSC data showed that the ability of PCL to crys-

Structure 14

tallise decreased with increasing copolyester content. The more compatible mixtures had the least tendency to crystallise; crystallinity was evident as ringbanded spherulites of the PCL component.

The possibility of transesterification in copolyester blends, with copolymer breaking PCL chains, was considered; such a process could modify the crystallisability and T_gs of blends. After heating blends to 260 °C for some time the T_gs determined from DSC cooling curves fell and then became constant. The effects were attributed to transesterification and results indicated that transesterification occurred fastest in blends containing copolymer with 40 wt % caprolactone and slowest in blends with copolymer having 61 wt % caprolactone. Heating also affected the ring-banded structure of the spherulites formed in the blends. From broad rings in the initial blends, heating resulted in thinner ring-banded structures and, on further heating, smaller spherulites [124].

Overall, it was concluded that the tendency for transesterification in the blends was a combination of a tendency for more rapid transesterification in blends containing copolymers with high butylene terephthalate, offset by a reduced miscibility in the blends containing copolymers with high butylene terephthalate content; reduced miscibility gave less contacts between potential reaction sites. The finer ring-banded structure of the spherulites which occurred in the earlier stages of reaction was attributed to the existence of copolymers of different composition in the blends; components with high and low butylene terephthalate contents behaved differently. The high butylene terephthalate copolymers were thought to react preferentially, to reduce the butylene terephthalate contents and enhance their miscibility with the remaining PCL [124].

15
Blends with Polycarbonates

15.1
Blends with Bisphenol-A Polycarbonate

Bisphenol-A polycarbonate 15 is normally amorphous (and clear) but is subject to solvent-induced crystallisation in the presence of some low-molecular-weight solvents [125].

Cruz et al. [21] established the general features of blends of bisphenol-A polycarbonate (PC) (Lexan 310, $\overline{M}_w$=29,200) with PCL (PCL-700, Table 1). Samples were prepared by melt mixing at 260 °C for no longer than 8 min; polyesters and polycarbonates are susceptible to interchange reactions on prolonged heating at

Structure 15

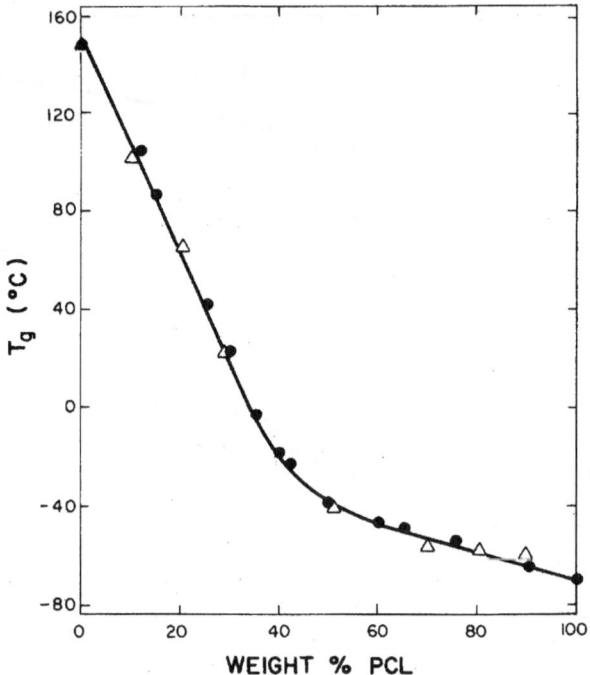

Fig. 42. Variations in glass-transition temperature in PCL/PC blends, (●) data from [21] and (Δ) [126]; taken from [126]

high temperatures and such processes could result in copolymer formation. Samples were subsequently melt-pressed at 250 °C. Pressed films were cooled in iced water and left to stand for 1 day (in the dry) at room temperature. Some samples were cast from mixed homogeneous solution in dichloromethane; solvent was removed rapidly at 50 °C and under vacuum to reduce solvent-induced crystallisation of PC.

Cruz et al. [21] demonstrated that all blends prepared as described above exhibited a single, composition-dependent T_g by differential thermal analysis (DTA), indicating miscibility in the amorphous phase. However, because of sample crystallinity there was no direct relation between T_g and overall composition. Glass-transition temperatures of similar samples, prepared by quenching from the melt, were studied by Jonza and Porter [126] and their data, together with those obtained by Cruz et al., are shown in Fig. 42. Blends containing more than about 30 wt % PCL exhibited a crystal melting endotherm for PCL and blends containing 30–90 wt % PCL also exhibited a high melting endotherm above 230 °C due to the presence of crystalline PC; crystal melting temperatures depended on the thermal history of the samples [21]. Recently, Hatzius et al. [127] also reported a single composition-dependent T_g and polycarbonate and PCL crystallinity over the same composition ranges in similar blends. They stat-

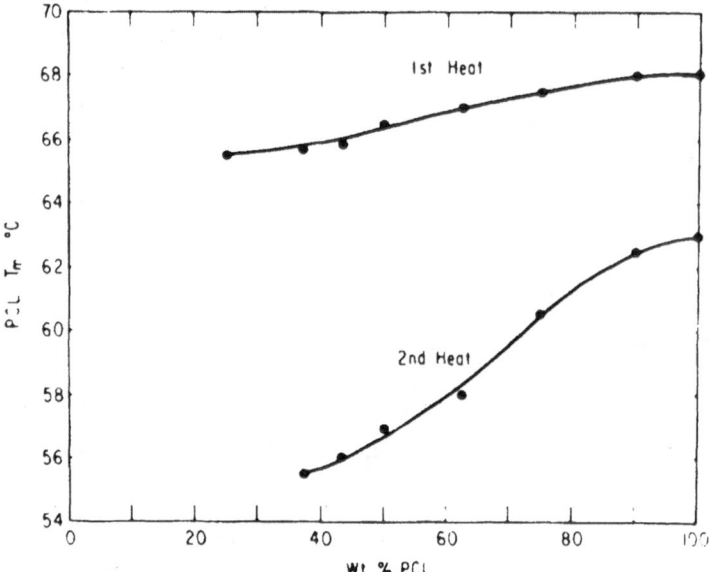

Fig. 43. PCL melting points in blends with PC; taken from [21]

ed that their limited T_g data, which differ from the preceding data in the mid-composition range, fit the Couchman-Karasz equation (Eq. 28) assuming a value of ΔC_p, the change in specific heat, for the glass transition of PCL (experimentally unobtainable due to crystallisation) of 0.43 ± 0.06 J g^{-1} and an experimentally determined value of 0.23 ± 0.02 J g^{-1} for polycarbonate. Thus, in blends of PCL and PC both polymers are often crystalline and the nature and thermal behaviour of the blends is complex:

$$\ln T_g = \frac{\Delta C_{pA}\phi_A \ln T_{gA} + \Delta C_{pB}\phi_B \ln T_{gB}}{\Delta C_{pA}\phi_A + \Delta C_{pB}\phi_B} \tag{28}$$

Samples, as prepared, were initially phase-separated (opaque) when hot but became homogeneous (clear) on cooling to room temperature, indicative of LCST behaviour. PCL crystallinity developed from the homogeneous melt and samples became cloudy again on further cooling. When examined by DTA, values of T_m for PCL were close to the normal melting points for PCL and decreased slightly with increasing PC content (Fig. 43). T_ms for PC also varied slightly (236–240 °C) with composition but in a more complex manner (Fig. 44); polycarbonate T_ms were lower than for pure polycarbonate (240–263 °C) [128]. After cooling at 10 °C min^{-1} and reheating, values of T_m for both PCL and PC were lower and samples containing only 20 wt % PCL no longer showed any crystallinity. Differences in behaviour reflected different thermal histories and probably different sizes of lamellae.

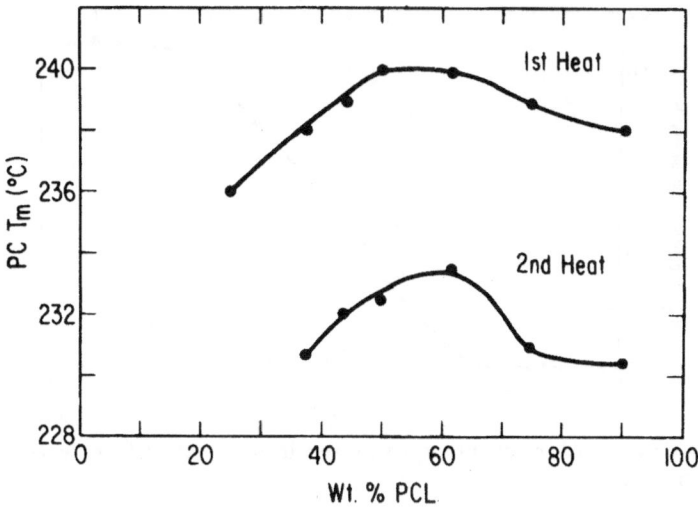

Fig. 44. PC melting points in PCL/PC blends; taken from [21]

In a separate study, Cruz et al. [129] attempted to investigate the basis of the miscibility in PCL/PC blends. They used low-molecular-weight analogues of various polyesters and polycarbonates. They determined heats of mixing and concluded that the heat of mixing of PCL and PC was negative. However, they did not identify the nature of the interaction between the constituents but simply suggested the existence of an interaction between the PCL carbonyl and the aromatic rings of PC. Jonza and Porter carried out a further detailed study of melting points of crystalline units in blends of PCL and PC in an attempt to assess interaction parameters from melting point depressions using the Nishi-Wang equation (Eq. 9) [126]. Hoffman-Weeks plots for blends containing 60–100% PCL all gave the same value of $T_{m,b}{}^0$, the equilibrium melting temperature for the blend, of 71±2 °C, which is equal to the equilibrium melting point for pure PCL (T_m^0) of 71 °C. The results indicated a value of the interaction parameter χ of zero or a slightly positive value; for the polymers used the critical value of χ (Eq. 6) for miscibility was 0.024 and lack of phase separation indicated that this value was not exceeded [126].

Ketelaars et al. recently examined the densities of amorphous blends (low PCL contents) and found that, while T_g data fitted Eq. (23), the densities of the samples were greater than would correspond to a weighted average [119]. They also showed that, at higher PCL contents, secondary crystallisation did not lead to an increase in density. These workers concluded that a single excess density parameter, indicative of favourable interactions between the components, is present in the amorphous material throughout the composition range.

Jonza and Porter could not determine values of equilibrium melting points for PC in the blends from Hoffman-Weeks plots [126] because the plots were

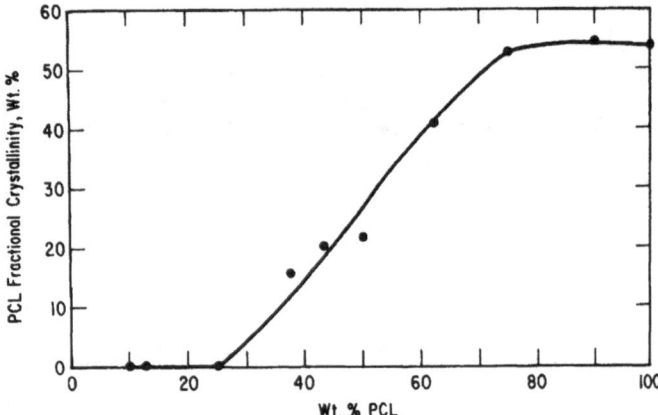

Fig. 45. Variation in PCL crystallinity (expressed in terms of PCL content) in PCL/PC blends; taken from [21]

non-linear. However, melting endotherms were composition independent, indicating that $\chi=0$. Non-linearity of the Hoffman-Weeks plots was attributed to a chemical reaction occurring between the components at the high temperatures involved. In contrast, Hatzius et al. [127] attributed the non-linear Hoffman-Weeks plots to changes in composition at the crystal growth front. They considered that, under different conditions of overall composition and temperature, the component not being incorporated into crystal locally might or might not have been able to diffuse away from the growth front and might, therefore, have caused changes in composition at the growth front. Such composition variations were considered to be responsible for the deviations from linearity and depression in the melting points observed at intermediate crystallisation temperatures where the local concentration of the non-crystallising component builds up; at higher crystallisation temperatures this component can diffuse away more rapidly. These authors used limiting points from the Hoffman-Weeks plots (at the lowest and highest crystallisation temperatures used) to estimate values of the equilibrium melting points. These values were put into the Nishi-Wang equation (Eq. 9), from which a value of the interaction parameter (-0.09 ± 0.01) and a crystal melting point for polycarbonate (248 ± 1 °C) were determined [127].

Cruz et al. [21] estimated crystallinities of samples from DTA data. Figure 45 shows how the crystalline fraction of PCL increased with overall PCL content from zero at about 25 wt % PCL to reach a plateau at about 80 wt %, comparable with that found in bulk PCL. Jonza and Porter obtained similar data [126]; PCL crystallisation increased from about 8%, in 50:50 (w/w) blends, to about 60% in blends with 70% PCL. These latter workers also studied PCL crystallisation kinetics and found equilibrium crystallisation was achieved (at 37 °C) in 10 min or less for samples containing 70% or more PCL. Crystallisation times in excess of 40 min were required for samples with less PCL (Fig. 46) [126]. Similar measurements indicated that PC crystallised over the same composition range, i.e. PC

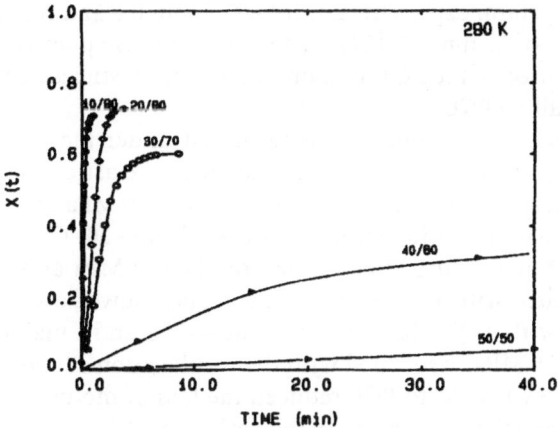

Fig. 46. Development of PCL crystallinity (X_t) in PCL/PC blends at 310 K; taken from [126]

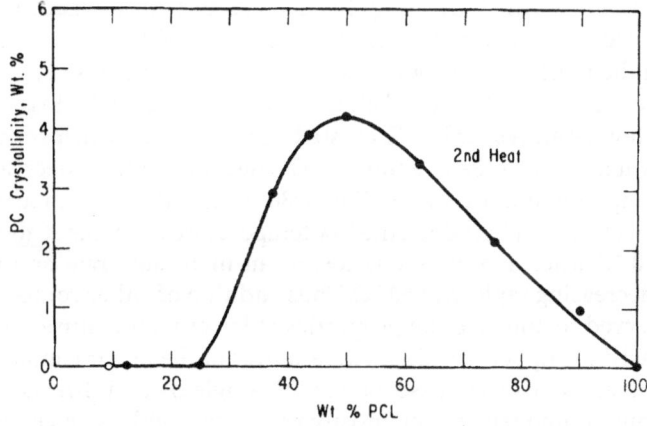

Fig. 47. PC crystallinity in PCL/PC blends based on total sample; taken from [21]

only crystallised when PCL crystallised. PC crystallinity was low under all conditions. The fraction of PC which crystallised was highest in blends containing 30–50 wt % PCL and only reached about 8%, according to Cruz et al. [21]; the PC crystallinity in blends (expressed as PC crystallinity in the total blend, PC plus PCL) is shown in Fig. 47. However, Hatzius et al., using an enthalpy of melting for PC of 148 J g^{-1} [130], calculated that PC crystallinity could reach 25% in blends containing 40 wt % PCL and crystallised at 160 °C, down to 12% for samples crystallised at 180 °C; crystallinities decreased rapidly with increasing PCL content [127]. Hatzius et al. further reported that PC crystallisation proceeded with a constant nucleation density and a growth rate of 10 mm h^{-1}; the crystallisation temperature was unspecified. It was also reported that for blends with

20 wt % PC the Avrami exponent (Eq. 26) was about 0.8 and constant $k\sim5.3$, giving a growth rate of 15 mm h^{-1} [127]. These workers also pointed out that the induced crystallisation of the PC is tantamount to antiplasticization of the polycarbonate by crystalline PCL.

Dynamic mechanical properties, determined under tension, were complex [21]. The major features for PCL were a small loss of elastic modulus associated with the β-relaxation, a decrease in modulus (from 20 MPa to about 6 MPa) associated with the α-relaxation (from about −60 °C to −20 °C), a continuing loss in modulus with increasing temperature, to about 4 MPa and a major loss in modulus associated with T_m from about 50 °C. PC showed a small loss in modulus associated with its β-relaxation at about −75 °C and a major loss of modulus, from about 15 MPa, associated with the α-relaxation at about 150 °C [21].

Addition of 25 wt % PC to PCL reduced the loss of modulus (and the size of the loss peak) associated with the α-relaxation of PCL, maintaining a slightly higher modulus at low temperatures (about 7 MPa at 0 °C) but reduced the temperature at which modulus was lost as a result of PCL melting to about 20 °C. The addition of 50% PC continued the trends to give a modulus of about 10 MPa at about 25 °C above which the modulus decreased. Thus the addition of PC to PCL only served to enhance mechanical properties of PCL at lower temperatures and reduced the mechanical properties at higher temperatures [21].

Addition of 12.5 wt % PCL to PC effectively counteracted the loss of modulus in PC at low temperatures [21]; PCL crystallinity presumably helped to maintain the modulus at about 20 MPa to about 50 °C. Thus the modulus decreased slowly with increasing temperature to 10 MPa at 80 °C and then fell catastrophically. However, 25 wt % PCL maintained the low temperature modulus at about 20 MPa but only to 0 °C when modulus was lost at an increasing rate and was about 1 MPa, and decreasing rapidly, at 50 °C. Thus addition of either component to the other only served to enhance the properties at lower temperatures and, in both cases, decreased the upper working temperatures of the major component [21].

All individual blends were associated with single α- and β-relaxations from the amorphous component, which mostly varied smoothly with changing composition [21].

All the PCL/PC blends with 20–80 wt % PCL were crystalline and cloudy at room temperature. The blends became optically clear when heated (to about 60 °C) and the PCL melted. They then became cloudy again at temperatures of about 260 °C with the onset of liquid-liquid phase separation occurring as a result of LCST behaviour. For PCL-700 (Table 1) the cloud-point curve was fairly flat over much of the composition range but rose slightly (about 5 °C) in blends containing 80 wt % PCL [21]. In comparable experiments with PCL-300 (Table 1) the cloud point curve varied more strongly with composition and was raised with a critical point at about 272 °C. These observations were consistent with the normal enhanced miscibility of lower-molecular-weight polymers.

We have already noted that PC can crystallise in the presence of PCL (which acts like a solvent) and Cruz et al. [21] reported that homogeneous samples quenched from a melt (during sample preparation) became cloudy at about T_g

as PCL crystallised. The blends then became clear when PCL melted (about 60 °C) but again became cloudy as PC crystallised from the melt, to become clear when PC melted (about 230 °C) and cloudy again when liquid-liquid phase separation occurred. Cruz et al. [21] reported that in their study there was no evidence of interchange reactions and that the appearance of a single T_g in the blends was probably a result of true miscibility in the amorphous phase. Jonza and Porter [126], however, reported evidence for a chemical reaction between the components when heated to 250 °C. Solubility studies indicated that after heating a 50:50 blend for 15 min only 15% of the PCL remained soluble and unreacted. Prolonged heating gave network formation. As a result of spectroscopic studies of heated blends, Jonza and Porter favoured the view that reaction was a thermo-oxidative branching process rather than transesterification.

Shuster et al. [131] studied interchange (transesterification) reactions. They prepared samples by melt mixing samples in a Brabender Plastograph and single-screw extruder at 240–260 °C. (They gave references to 16 papers on transesterification in polycarbonate plus polyester systems in which reactions are often catalysed by titanium compounds.) They noted the absence of previous evidence of transesterification in uncatalysed blends. Interchange reactions were expected to occur by acidolysis between carbonyl ends of PCL and PC and by transesterification between ester and carbonate. The first step in both reactions would be heterolysis of -CO-O-; all reactions might be catalysed. The acidolysis catalysts investigated were p-toluenesulfonic acid and adipic acid. Esterolysis catalyst studied was dimethylterephthalate and transesterification catalysts were organo-titanium and organo-tin compounds. Polymers were first melt mixed (240 °C) and catalyst was added after some time. No changes were observed as a result of adding adipic acid, dimethylterephthalate and triphenyl phosphite and it was concluded that no interchange occurred.

In acid-catalysed processes, the addition of p-toluene sulfonic acid was found to cause a reduction in torque in the mixer, darkening, a loss of melt viscosity and enhanced crystallisability of the PC in the presence of 10–20% PCL. Shuster et al. [131] suggested that the enhanced crystallinity meant that there was no transesterification and p-toluene sulfonic acid promoted carbonate+carbonate and not carbonate+ester interchange. Previously Smith et al. [132] had reported that transesterification produced block copolymer and loss of crystallisation and crystallisability.

When adding titanium compounds Shuster et al. [131] found that Ti(OBu)$_4$ and Ti(OEt-NH-Et-NH$_2$)$_3$ caused torque reductions in blends with 10–30 wt % PCL, discoloration (white to red), gas evolution and foaming of the molten blend; effects were more pronounced for Ti(OBu)$_4$. In studying the Ti(OBu)$_4$ system, Shuster et al. [131] noted that the uncatalysed physical blend with 25% PCL gave PC T_m at 220 °C and $T_g\sim$30 °C; i.e. PCL plasticized PC and promoted crystallisation. Addition of 0.1% Ti(OBu)$_4$ reduced the melting endotherm, broadened it and shifted it to lower temperatures ($\sim$210 °C) and broadened T_g. Addition of 0.2–0.3% Ti(OBu)$_4$, caused the melting point to disappear and raised and sharpened T_g; with 25% PCL the blend was transparent, but orange red; the original blend was opaque. Similarly in 30% PCL blend, the PC crystallinity decreased; the original sample showed two T_ms, one each from PCL and PC. However, 0.1% Ti(OBu)$_4$ changed this picture; it removed the PCL T_m and shifted PC T_m; 0.2–0.3% Ti(OBu)$_4$ gave one T_g. The several changes were attributed to copolymer formation, reduction in crystallinity and heterogeneity in the amorphous phase (as evidenced by a broad T_g) and then to further randomisation with more catalyst. Polymer solubility changed on reaction in the presence of Ti(OBu)$_4$ (some PCL became insoluble in CCl$_4$) and the infrared spectrum changed; there was no gel formation.

Ti(OBu)$_4$ is moisture sensitive and gives rise to discoloration. Therefore, it was suggested, tin compounds might be preferable as practical catalysts, even though they are less effective as exchange catalysts but gave no discoloration or foaming. For melt processing in an extruder shorter residence times can be used and catalysed reactions may take place in preference to thermal ones [131].

15.2
Blends with Substituted Polycarbonates

The study of blends of PC with PCL ($\overline{M}_n$=15,000, $\overline{M}_w$=46,700) and other polyesters was extended to tetramethyl bisphenol-A polycarbonate (TMPC) 16 by Fernandes et al. [86]. The sample of TMPC used (η_{rel}=1.29) had T_g=193.5 °C and T_m≈280 °C. Samples were cast from toluene or dichloromethane and, when relatively dry from solvent, residual solvent was removed under vacuum at 80 °C (dichloromethane) or 110 °C (toluene).

The thermal transition behaviour of these blends was studied by DSC, usually taking data on a first heating cycle. As with PC blends, the thermal transition behaviour was somewhat complex; toluene was reported to induce crystallinity in TMPC (Fig. 48). For blends cast from toluene, samples showed a single composition-dependent T_g. Samples containing at least 60 wt % PCL also exhibited PCL and TMPC crystallinity; a very broad transition was observed in blends with 60 wt % PCL. When cast from dichloromethane, blends rich in PCL exhibited a single T_g but blends rich in TMPC exhibited two T_gs. The difference in behaviour it was suggested [86], following the arguments of Robard et al. [133], was that toluene had similar interaction parameters with both polymers, giving a "solvent effect" which induced miscibility.

Fernandes et al. [86] also determined enthalpies of fusion for the blends cast from both solvents (Fig. 49). Enthalpies of fusion for PCL were the same in both blends but differences were recorded for TMPC melting. Enthalpies of fusion

Structure 16

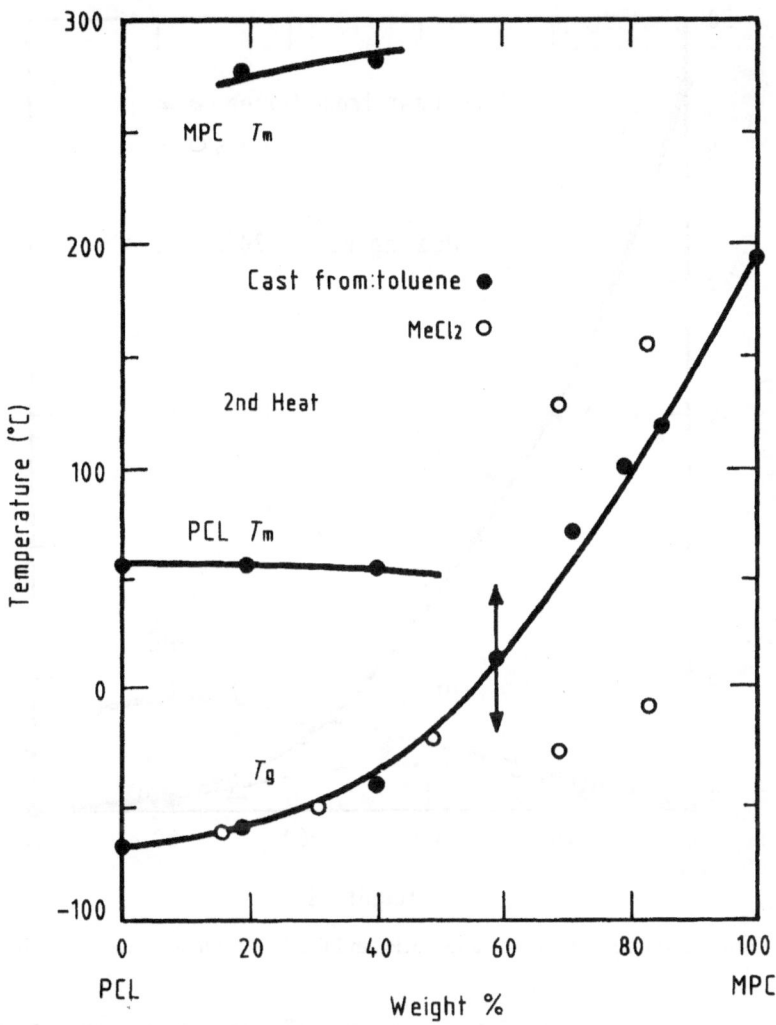

Fig. 48. Thermal transition temperatures in PCL/TMPC blends; taken from [86]

were identified and determined for both sets of blends at virtually all compositions although values of T_m for TMPC were only recorded in certain cases (Fig. 49). The authors speculated that TMPC crystallinity might be induced by toluene and by PCL.

Because PC and TMPC form miscible blends in all proportions at all temperatures, there are problems in determining their mutual interaction parameters but a value for B of -7.5 J cm^{-3} was estimated [86]. Kim and Paul [87] examined ternary blends of PCL with PC and TMPC to provide an indirect method of estimating interaction parameters. Samples of the ternary blends of PCL (PCL-

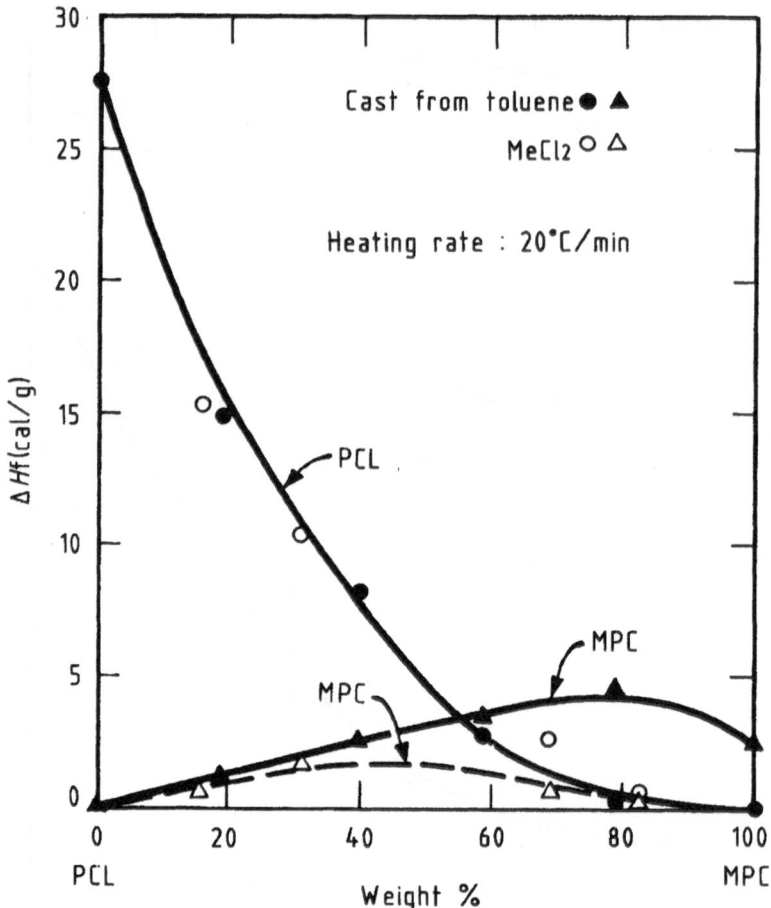

Fig. 49. Enthalpies of fusion for PCL and TMPC in PCL/TMPC blends; taken from [86]

700, Table 1), PC (Lexan 131–11, $\overline{M}_w$=38,000, $\overline{M}_n$=13,000) and TMPC ($\overline{M}_w$= 33,000) were solvent cast from THF, solvent was evaporated at 80 °C and then under vacuum at 80 °C for one week and samples were quenched to room temperature. The value of the interaction parameter B determined was –8.2 J cm^{-3} [86].

Thermal analysis by DSC showed single, composition-dependent T_gs at all PCL contents in blends with both PC/TMPC 1:1 and PC/TMPC 7:3 (Fig. 50). Kim and Paul stated that this implied that blends were miscible in all proportions. However, the blends were cloudy at room temperature due primarily to PCL crystallinity and observed T_gs were not necessarily representative of the overall composition as amorphous polymer. The cloudiness decreased above T_m of TMPC; no liquid-liquid phase separation was observed up to 350 °C, and the systems were totally miscible at high temperatures.

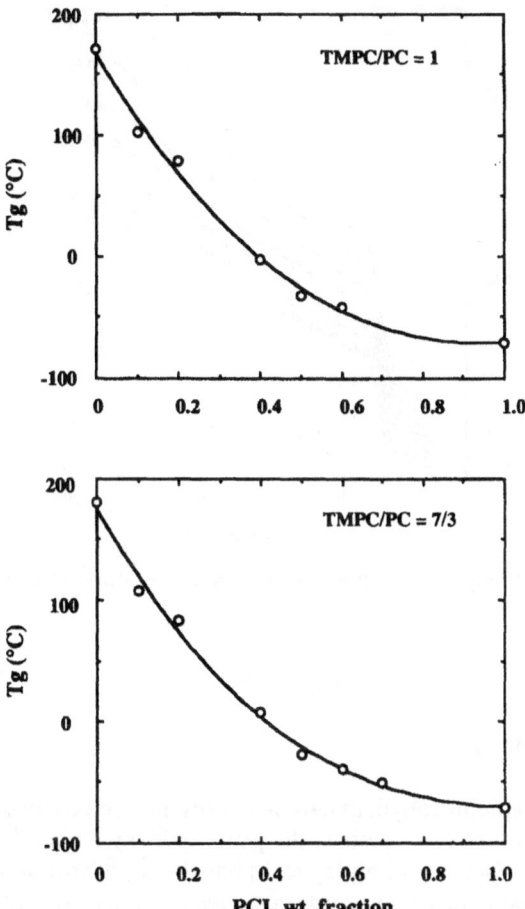

Fig. 50. Glass-transition temperatures (onset) for PCL/TMPC blends determined by DSC at a heating rate of 20° C min⁻¹; taken from [87]

Estimates of equilibrium melting temperatures of PCL crystallites were made, for various binary and ternary systems, with the aid of Hoffman-Weeks plots. The observed melting points of PCL were between 55 °C and 65 °C and varied by about 3 °C with changes in T_c from 38 °C to 48 °C. From the melting-point depressions, as a function of composition and pressure-volume temperature data, Kim and Paul estimated equation-of-state parameters. During the course of this study Kim and Paul determined specific volume data for PCL at a series of hydrostatic pressures (Fig. 51) [87].

Dynamic mechanical analysis data for ternary blends gave a similar pattern of behaviour as for the binary blends of PCL/PC already discussed. The variation in T_g with composition determined from the α-relaxations closely mirrored that from DSC data.

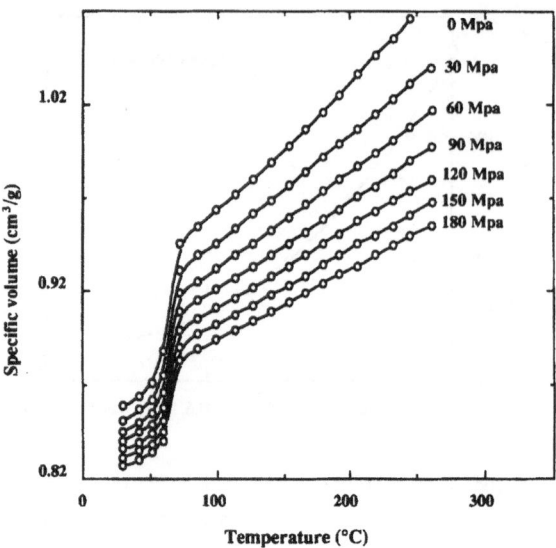

Fig. 51. Variation of the specific volume of PCL with temperature at various pressures; taken from [87]

16
Blends with Phenoxy

Bisphenol-A and epichlorohydrin can be copolymerised to form a regular polymer with the structure **17** known as the poly(hydroxy ether of bisphenol-A) or, more generally, as phenoxy. Phenoxy resin has the same aromatic unit as bisphenol-A polycarbonate but, as far as interactions are concerned, the carbonate residue is replaced by an ether linkage and a hydroxyl group; the latter is possibly capable of useful interactions.

Brode and Koleske [134] demonstrated that PCL forms compatible (miscible) blends with phenoxy. They melt-mixed the components and prepared plaques by compression moulding. Samples with more than about 50 wt % PCL ($\overline{M}_w$= 44,000) exhibited PCL crystallinity; crystallisation took place over a few days at room temperature during which period samples became hazy and the modulus increased. Quenched samples were used to determine the glass-transition temperature behaviour of amorphous blends. Samples showed a single composition-dependent T_g, indicating miscibility in the amorphous material; values of

Structure 17

T_g fitted well to the calculated curve based on Eq. (23), using a value of -71 °C for the T_g of PCL.

Blends of PCL (PCL-700, Table 1) with phenoxy resin (PKHH, $\overline{M}_w$=80,000, $\overline{M}_n$=23,000) were also studied by Harris et al. [83]. Samples were prepared by solvent casting from THF at room temperature followed by drying under vacuum for 24 h at 100 °C. Glass transition temperatures of PCL/phenoxy blends determined by DSC after thermal cycling above and below limiting transition temperatures agreed well with previous data reported by Brode and Koleske [134]. Values of the crystal melting temperatures of PCL, in blends containing up to 60 wt % phenoxy, were also recorded. Estimated heats of fusion per unit mass of total sample indicated that the extent of PCL crystallisation in the blend was less than that in pure PCL. Harris et al. [83] argued that this result is probably because of a lower rate of crystallisation caused by an increase in T_g of the blend by adding phenoxy (T_g~98 °C).

From melting point depressions Harris et al. [83] estimated a value of the interaction parameter B of -10.1 J cm^{-3}. These authors suggested that the interactions responsible for miscibility involve hydrogen bonding with the carbonyl of PCL and the hydroxyl of the phenoxy, probably counteracted to some extent by endothermic interactions between other structural units.

Coleman and Moskala [85] investigated interactions in PCL/phenoxy blends by FTIR, employing samples cast from THF solution as thin films on KBr discs. These workers identified changes in the FTIR spectra of phenoxy with temperature which related to changes in concentration of free and H-bonded hydroxyl groups; they had previously investigated PCL spectra and identified crystalline and amorphous carbonyl-stretching peaks. From studies of spectra of blends containing different proportions of phenoxy at 75 °C, where the blends are homogeneous and amorphous, Coleman and Moskala noted the appearance of a new peak attributable to a carbonyl hydrogen-bonded to phenoxy. They estimated the fraction of H-bonded PCL carbonyl groups and found that the fraction increased with phenoxy content; the concentration of pure hydroxyls in phenoxy did not vary much. They estimated that a significant fraction of phenoxy hydroxyls, possibly 0.6, was associated with PCL carbonyl.

Significantly, Coleman and Moskala concluded that the strength of the PCL-phenoxy H-bonding in the blend is weaker than that in phenoxy itself; this indicates a positive enthalpy of mixing in PCL-phenoxy blends. Nevertheless, the two polymers form miscible blends and Coleman and Moskala concluded that, for the overall free energy of mixing to be negative, miscibility must be driven largely by the entropy of mixing. Normally the entropy of mixing for polymer pairs is trivial but increases with decreasing molecular weights of the components and PCL-700, used in these studies, is of modest molecular weight.

For semicrystalline polymers the same workers identified relevant absorptions for crystalline and amorphous bands and concluded that, for blends containing 70 wt % or more phenoxy, there was no PCL crystallinity and blends were totally amorphous [85]. This limitation is consistent with the data of Harris et al. [83].

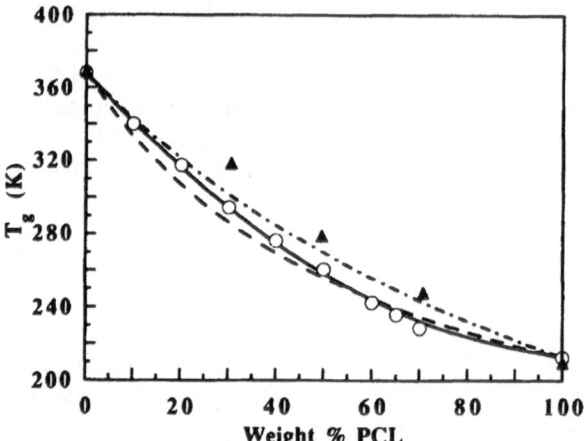

Fig. 52. Glass-transition temperature variations in PCL/phenoxy blends: (○) data from [137] and (▲) from [59], lines calculated from theoretical equations; taken from [137]

Espi and Iruin made estimates of phenoxy miscibility with several polyesters using a group contribution method [135]. Their data were compared with heats of mixing of model compounds and, on that basis, they concluded that PCL/phenoxy blends have negative enthalpies of mixing, which contradicts the conclusions of Coleman and Moskala [85].

The miscibility of PCL with phenoxy and the crystallisation and melting behaviour of the blends was also studied by Defieuw et al. [59]. They coprecipitated PCL ($\overline{M}_w$=22,500; $\overline{M}_w/\overline{M}_n$=1.60) and phenoxy ($\overline{M}_w$=48,600; $\overline{M}_w/\overline{M}_n$=2.17) from THF solution into hexane; samples were then compression moulded at 120 °C and quenched into liquid nitrogen in order to obtain amorphous blends. Dynamic mechanical analysis of quenched blends, both above and below the PCL T_m, showed narrow tanδ peaks characteristic of miscible (homogeneous) materials. T_gs showed a single composition-dependent T_g and were said to be in good agreement with Eq. (23) but lie above the calculated line. Crystallisation proceeded with a constant rate of radial growth at 45 °C, which decreased with increasing phenoxy content, as PCL crystallised from the blend in a liquid-solid phase separation. This observation suggested intraspherulitic segregation of the phenoxy as PCL spherulites grew at PCL contents greater than 60 wt %. At higher phenoxy concentrations the spherulites were non-space filling and some phenoxy was located interspherulitically.

Defieuw et al. [59] also undertook SAXS studies and found that the long spacing in the spherulites varied with phenoxy content and exhibited a maximum at about 20 wt % phenoxy. This is a similar result to that previously found in blends of PCL with chlorinated polyethylene [100]; phenoxy demonstrated a greater tendency to segregate interlamellarly than CPE although it has a similar T_g. Defieuw et al. discussed the origin of the differences in segregation behaviour. Vanneste et al. [136] also confirmed previous studies that in PCL/phenoxy binary blends the amorphous layer thickness is less than in pure PCL and suggested that strong interactions between the components, in the form of hydrogen-bonding, could be responsible for the interfibrillar location of the amorphous material, between stacks of lamellae, rather than interfibrillar location of amorphous PCL between lamellae as found in pure PCL [60].

It was also observed [59] that, on melting, double endotherms were observable by differential scanning calorimetry; the temperature corresponding to the peak of the upper melting peak decreased with increasing phenoxy content (melting point depression). No melting endotherms were observed in phenoxy-rich (>50 wt %) blends and lack of crystallinity was attributed to the high T_g of the blends, compared with the crystallisation temperature. The lower melting endotherm appeared in samples crystallised for long times and was attributed to secondary crystallisation.

De Juana and coworkers carried out a series of studies on blends of phenoxy ($\overline{M}_w$=50,700, $\overline{M}_n$=18,000) with PCL ($\overline{M}_w$=17,600, $\overline{M}_n$=10,800) [84, 137–139]. They confirmed the existence of a single glass-transition temperature in quenched samples by DSC [137] as previously observed by others [59, 134]. The combined data from [59] and [137] are shown in Fig. 52 which shows that the data are in general accord with the predictions of the Gordon-Taylor equation (Eq. 22), confirming miscibility of the polymers in all proportions in the melt. Glass-transition temperatures in annealed samples, determined from the α-relaxation peak from dynamic mechanical analysis data (1 Hz) were considerably higher than values determined by DSC in samples containing <70 wt % phenoxy, reflecting the presence of crystalline PCL and higher phenoxy contents in the amorphous component, compared with the overall composition. For samples with >70 wt % phenoxy, in which PCL crystallisation was inhibited, glass-transition temperatures by DSC and dynamic mechanical analysis were in approximate agreement. These data confirmed the existence of a single amorphous phase in the PCL-phenoxy blends.

In contrast, dynamic mechanical analysis data showed the presence of two β-relaxation peaks which corresponded closely to, and were attributed to, the β-relaxations in the individual components (Fig. 53) [138]. These data were interpreted as being due to secondary motions responsible for the β-relaxations in each component which were largely unaffected in the blends; e.g. motions in the $(CH_2)_n$ sequences (n>4) at about –110 °C in PCL were unaffected by the presence of the phenoxy. In detail, it was shown that T_β (the temperature of the maximum of the β-relaxation peak) for phenoxy decreased with increasing PCL content (Fig. 54) [138]. This effect was attributed to H-bonding of the phenoxy -OH to the carbonyl of the PCL restricting molecular motions in the phenoxy.

The kinetics of isothermal crystallisation were studied in quenched (at –120 °C) amorphous samples after heating to 127 °C and rapid cooling to the crystallisation temperature [137]. Crystallisation kinetics, for PCL crystallisation, were consistent with the Avrami equation (Eq. 26) with values of n close to 3, consistent with spontaneous heterogeneous nucleation and three-dimensional growth of spherulites, at crystallisation temperatures from 30 °C to 40 °C. Rates of crys-

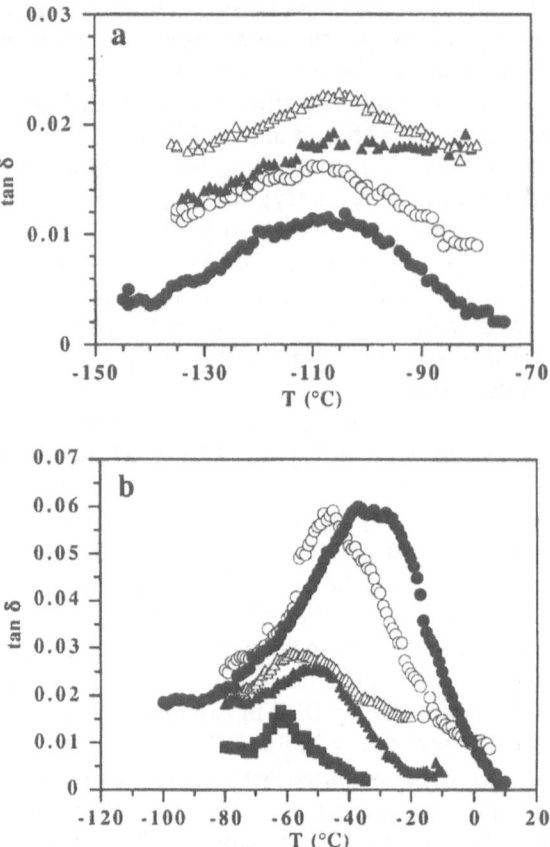

Fig. 53a,b. β-Relaxations in PCL/phenoxy blends for: **a** PCL relaxations; **b** phenoxy relaxations for different compositions of the blends; taken from [138]

tallisation decreased with increasing phenoxy content; times for 100% PCL crystallisation increased from about 100 s to 1200 s as the phenoxy content increased from zero to 35 wt %. The kinetic data, expressed in terms of $t_{1/2}$, the half-time for crystallisation given by

$$t_{1/2} = \left(\frac{\ln 2^{1/n}}{k} \right)$$

are given in Fig. 55 [137].

Melting point depressions in isothermally crystallised blends were analysed in terms of the Nishi-Wang equation (Eq. 10) [40] using equilibrium values of the melting point determined from Hoffman-Weeks plots; melting points were determined by DSC and optical microscopy [137]. Values of the melting-point

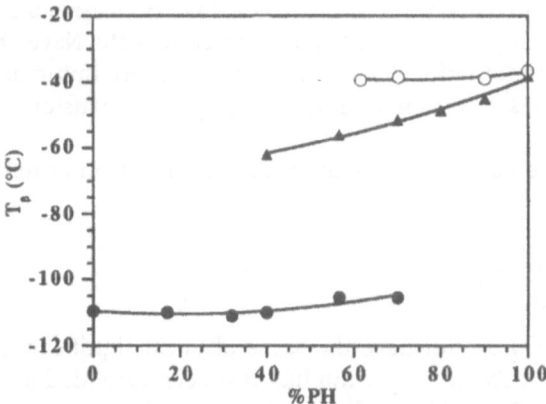

Fig. 54. Variations in β-relaxation maxima for PCL (●) and phenoxy (▲) in PCL/phenoxy blends; adapted from [138]

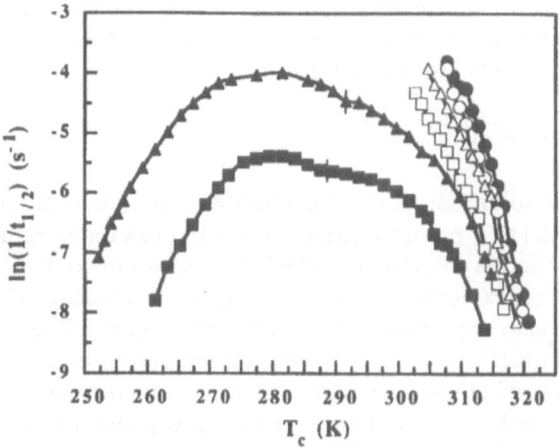

Fig. 55. Kinetic data for PCL crystallisation in PCL/phenoxy blend. ● 100% PCL, ■ 60% PCL taken from [137]

depressions were small and led to an estimate of the interaction parameter B of about −10.1 J cm⁻³ (by microscopy) and −9.0 J cm⁻³ by DSC, compared with a value of −10.1 J cm⁻³ determined by Harris et al. [83].

De Juana et al. [84] also used inverse-phase gas chromatography to investigate interaction parameters χ_{AB} for each component with a series of common solvents at 150 °C. They also determined B for PCL-phenoxy blends by different techniques for several compositions and temperatures. Values determined varied from −4.8 J cm⁻³ to −16.1 J cm⁻³.

In a separate study, de Juana et al. investigated non-isothermal crystallisation in PCL-phenoxy ($\overline{M}_w$=50,700) blends [139]. The detailed results of such studies inevitably reflected rates of cooling and heat loss. Hence, results in general were influenced by such parameters as sample thickness etc. Nevertheless, non-iso-thermal crystallisation kinetics are, in principle, more immediately relevant to practical situations such as injection moulding and extrusion than are isother-mal experiments.

The Avrami equation (Eq. 26), applicable to non-isothermal crystallisation, can be written as

$$X(t) = 1 - \exp\left(-K_n t^n\right) \tag{29}$$

in which K_n and n do not have the same physical significance as K and n in Eq. (26) [140]. The Avrami equation has also been extended to non-isothermal crystallisation by Ozawa [141] who obtained the relation

$$\log\left\{-\ln[1 - X(t)]\right\} = \log K(t) - n\log\beta \tag{30}$$

where β is the cooling rate and $K(t)$ is a cooling function; $X(t)$ has the same meaning as in (Eq. 26). A third approach is to use another equation due to Zia-biki [142, 143], Eq. (31), similar to Eq. (26):

$$X(t) = 1 - \exp\left[-E(t)\right] \tag{31}$$

where $E(t)$ is the volume fraction of crystals nucleated at some time s, s<t.

De Juana et al. [139] reported the variation in peak temperatures of the crys-tallisation endotherms (obtained by DSC) for a series of cooling rates for blends with different phenoxy contents (Fig. 56). They also recorded the widths of the crystallisation peaks expressed in terms of time (Fig. 57). These data show that, in PCL-rich blends (PCL>60 wt %), the peak temperatures and peak widths de-crease with increasing cooling rates; starting temperatures for cooling were 127 °C (400 K). Peak temperatures were between about 17 °C and 32 °C, peak widths varied from 2 min to more than 30 min (at a cooling rate of 0.3 °C min^{-1}). Peak temperatures were lower and widths higher at the higher phenoxy con-tents. These observations are consistent with reduced rates of crystallisation in the presence of the higher-viscosity phenoxy and increasing phenoxy contents in the amorphous phase as crystallisation of PCL proceeds.

Extents of crystallinity $X(t)$ were determined from integration of the exo-therms (Fig. 58); these curves were sigmoidal and very similar to those from the Avrami equation (Eq. 26). They showed that, for a given composition, crystalli-sation at lower cooling rates starts at higher temperatures, as might be expected. From plots of Eq. (30) values of n were close to 3 for PCL contents >70 wt %, fall-ing to 2.2 at PCL contents of 60 wt %; the higher values are consistent with the results from isothermal crystallisation.

Application of the modified Avrami equation (Eq. 29) gave values for n of about 4 (decreasing to 3.1 at 60 wt % PCL); these higher values were attributed

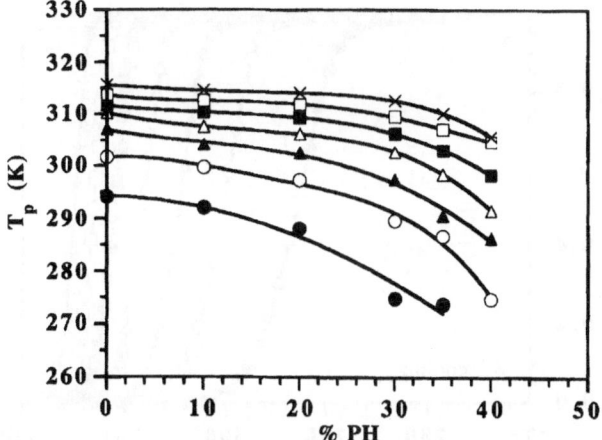

Fig. 56. Variations in temperature of crystallisation peaks in PCL/phenoxy blends at cooling rates: (●) 20 °C min⁻¹, (○) 10° C min⁻¹, (▲) 5° C min⁻¹, (Δ) 2.5° C min⁻¹, (■) 1.25° C min⁻¹, (□) 0.62° C min⁻¹, (✕) 0.31° C min⁻¹; taken from [139]

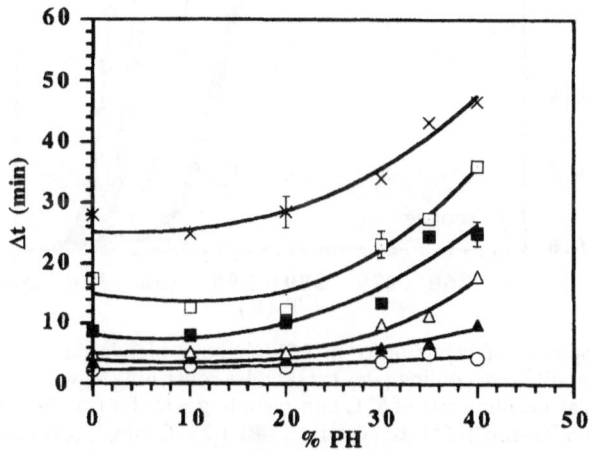

Fig. 57. Widths of crystallisation peaks, expressed in terms of time, for crystallisation in PCL/phenoxy blends, cooling rates: (●) 20° C min⁻¹, (○) 10° C min⁻¹, (▲) 5° C min⁻¹, (Δ) 2.5° C min⁻¹, (■) 1.25° C min⁻¹, (□) 0.62° C min⁻¹, (✕) 0.31 °C min⁻¹; taken from [139]

to changes in growth rates of the spherulites as temperatures decreased during cooling. Details of the analysis following the procedure of Ziabiki can be found in the original publication [139]. Although there was general agreement between the several theoretical approaches they all failed to give a complete description of non-isothermal crystallisation.

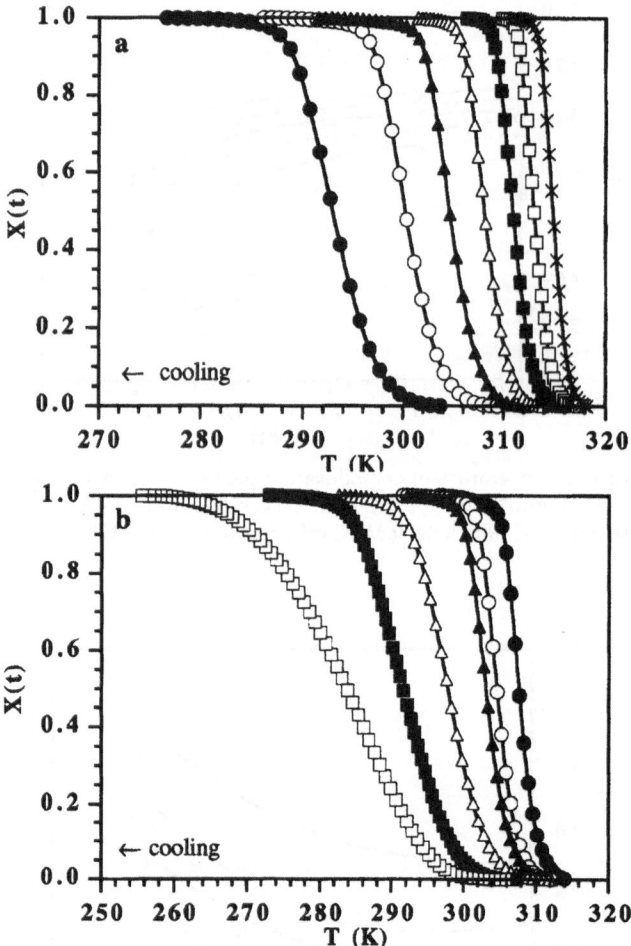

Fig. 58a,b. Development of crystallinity in PCL/phenoxy blends during non-isothermal crystallisation: **a** at different cooling rates for samples containing 90 wt % PCL: **b** for different compositions at a cooling rate of 5° C min⁻¹; cooling rates for (**a**): (●) 20° C min⁻¹, (○) 10° C min⁻¹, (▲) 5 °C min⁻¹, (△) 2.5 °C min⁻¹, (■) 1.25 °C min⁻¹, (□) 0.62 °C min⁻¹, (✕) 0.31 °C min⁻¹, compositions for b (wt % PCL): (●) 100, (○) 90, (▲), 80, (△) 70, (n)65, (□) 60; taken from [139]

17
Blends with Other Styrene-Containing Polymers

We have already considered blends of PCL with styrene-acrylonitrile copolymers (Sect. 11), which constitutes one of the major studies of PCL blends. In addition, there have been several smaller studies with polystyrene and other styrene-containing polymers which are considered here.

17.1
Blends with Polystyrene

High-molecular-weight polymers of different chemical structure are normally immiscible but as the molecular weights of the components are decreased (below degrees of polymerisation of about 100) the entropy of mixing and the driving force responsible for miscibility in low-molecular-weight substances increases rapidly. (Sect. 3.1) Polystyrene is normally immiscible with PCL but it has been reported that oligomeric polystyrene (oligo-PS) is miscible with PCL and that the system exhibits upper critical solution temperature behaviour, which is characteristic for many low-molecular-weight mixtures [144], quoted in [145]. However, Svoboda et al. [77] reported negative heats of mixing for polystyrene and PCL, based on microcalorimetric studies of low-molecular-weight analogues, which should imply miscibility of high-molecular-weight polymers; the reasons for immiscibility are unclear but possibilities were discussed (see Sect. 7) [77].

Blends of PCL with oligo-PS were studied by Tanaka and Nishi [22, 146] (PCL $\overline{M}_w$=33,000, $\overline{M}_n$=10,700, oligo-PS $\overline{M}_w$=950, $\overline{M}_n$=840), by Li et al. [81] (PCL-300, Table 1; oligo-PS $\overline{M}_w$=840, $\overline{M}_n$=765) and by Nojima et al. [145] (PCL $\overline{M}_w$= 13,700, $\overline{M}_w/\overline{M}_n$=1.44). Samples were obtained by casting mixtures from common solvents and heating to homogenise them prior to investigation. The three groups reported similar phase diagrams which showed a two-phase region bounded by a binodal at high volume fractions of oligo-PS; details of the phase diagrams, such as critical temperatures, varied strongly with the molecular weights of the constituents. Figure 59 shows the phase diagram reported by Li et

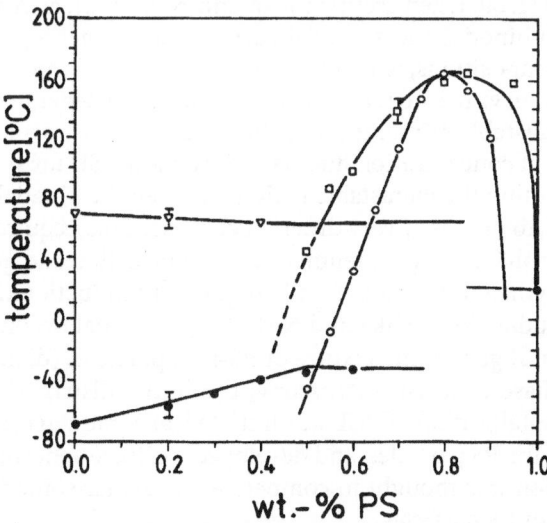

Fig. 59. Phase diagram for PCL/PS blends: ∇ homogeneous melt/PCL crystal coexistence curve, ● glass-transition temperatures, O spinodal and □ binodal; taken from [80]

al. [81] from which it is seen that the system is characterised by UCST behaviour and liquid-liquid phase separation of oligo-PS-rich mixtures to give a phase of almost pure oligo-PS and a mixed phase. At high PCL contents, where the system is miscible to relatively low temperatures, the crystal melting curve for PCL intersects the binodal.

Tanaka and Nishi [146] and Li et al. [81] interpreted experimental points on the binodal in terms of Flory-Huggins theory and determined values of interaction parameters and the binodal curves. Both groups noted positive values of the interaction parameters; values are quoted in Sect. 7. Li et al. [81] also calculated the spinodal in Fig. 59 from their value of $\chi_{1,2}$. Miscibility in this system, therefore, depends on a positive entropy of mixing and is restricted to low molecular weights of the constituents.

Li et al. [81] and Nojima et al. [145] noted very small equilibrium melting point depressions for PCL. The former, using equilibrium values of T_m from Hoffman-Weeks extrapolations, found a melting point depression of about -0.14 K (wt %)$^{-1}$ PS. Tanaka and Nishi [146] showed that the depressions were consistent with the Nishi-Wang equation (Eq. 9).

Li et al. [81] also quenched homogeneous samples rich in PCL in an attempt to determine T_gs for unstable homogeneous amorphous mixtures. Values of T_g determined on heating were found to be in good agreement with Eq. (23) for compositions outside the binodal.

Tanaka and Nishi identified several regimes for phase separation from homogeneous mixtures in which there might be competition between different processes leading to different morphologies [22]. These regimes are standard for any pair of polymers, one of which is crystallisable, which are miscible over a limited range of compositions as in Fig. 59; these regimes are identified in the schematic phase diagram, Fig. 60, taken from Tanaka and Nishi [22], as A to D. This diagram shows both binodal and spinodal curves and the melting point curve for homogeneous compositions, which has been extrapolated into the phase-separated region; this curve has no real meaning in the phase-separated region. Within the spinodal, unstable mixtures inevitably undergo phase separation through the development of concentration fluctuations (denoted SD for spinodal decomposition) while within the metastable region, between the binodal and spinodal, phase separation, to phases of very different compositions, requires phase nucleation followed by phase growth (denoted NG for nucleation and growth); for nucleation and growth, nucleation normally requires an induction time.

For Case A, studied by Tanaka and Nishi [22], spinodal decomposition (of a 40 wt % PCL blend) gave a fine texture of PS-rich particles (diameter ~10 µm), observable by phase-contrast microscopy; the initial distribution of particles was uniform. Crystallisation of PCL was initiated, at some stage, in the PCL-rich regions, between the PS particles, and developed a diffuse spherulite-type structure; crystallisation was thought to compete with further spinodal decomposition and growth of PS particles.

A second case studied by Tanaka and Nishi [22, 146] and also by Li et al. [81] was of a PCL-rich system (70 wt % PCL, regime C) in which phase separation of

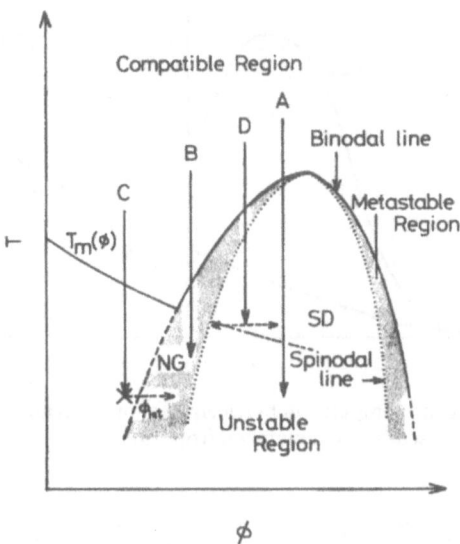

Fig. 60. Schematic phase diagram, relevant to PCL/PS blends, ϕ is PS content, to define different cooling regimes for discussion of phase separation and crystallization, as discussed in the text; taken from [22]

the initial blend did not occur because of thermodynamic immiscibility of the amorphous components into two amorphous phases but by crystallisation of the crystallisable component. Formation of crystalline PCL (100 wt % PCL) depletes the amorphous phase of PCL and changes its composition, locally, into a composition richer in PS and, ultimately, into the immiscible region of the phase diagram. Li et al. [81] identified three sub-regimes corresponding to different relative rates of crystallisation and diffusion (migration) of the PS-rich phase. Competition between these processes is shown schematically in Fig. 61 [81]. C(I) corresponds to very rapid crystallisation compared to diffusion, C(II) to the regime in which the processes are competitive and C(III) where diffusion/migration is rapid relative to crystallisation; these sub-regimes correspond to different crystallisation temperatures. Similar observations were made by the two groups but details differed.

At the lowest T_c, where the driving force for PCL crystallisation was highest, space-filling spherulites with the PS contained intraspherulitically were reported by Li et al. [81], whereas Tanaka and Nishi found interspherulitic impurity (PS-rich) layers [146].

At higher crystallisation temperatures (regime C(II)) spherulites developed droplets of PS-rich material at their growth surfaces (Fig. 62); according to Tanaka and Nishi [146] a small gap existed between the droplets and growth surfaces in the early stages (1860 min at 50 °C) which disappeared later as the droplets increased in size and became non-spherical (5000–6000 min at 50 °C). The par-

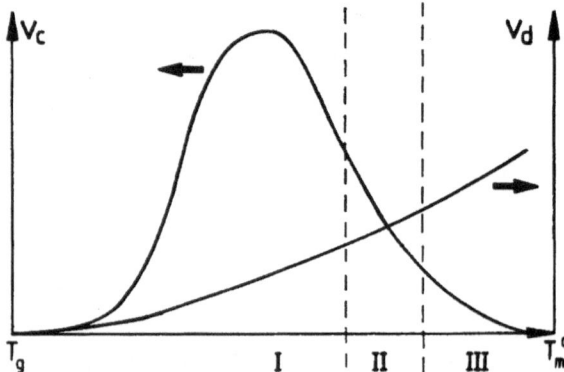

Fig. 61. Diagram to illustrate competition between rates of spherulite growth V_c and diffusive processes V_d, as discussed in the text; taken from [81]

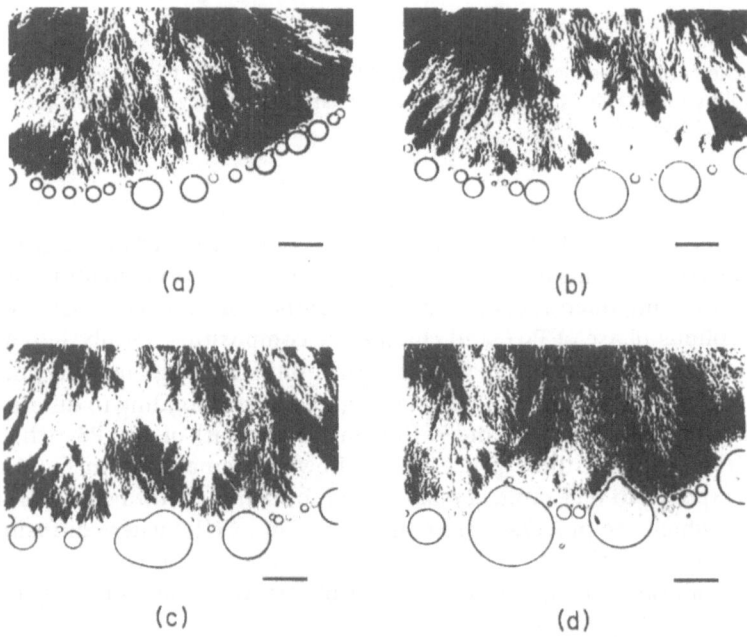

Fig. 62a–d. Morphologies near the growth surfaces of spherulites in PCL/PS blends showing the development of PS droplets between the PCL crystallites and the mixed melt phase; taken from [146]

ticles reduced the rate of growth of the spherulites and the impinging of growing spherulites caused the PS-droplets to be exuded and to coalesce. Li et al. [81] commented that PS released during crystallisation of the PCL diffused to the surrounding melt and, eventually, the composition of the surrounding melt

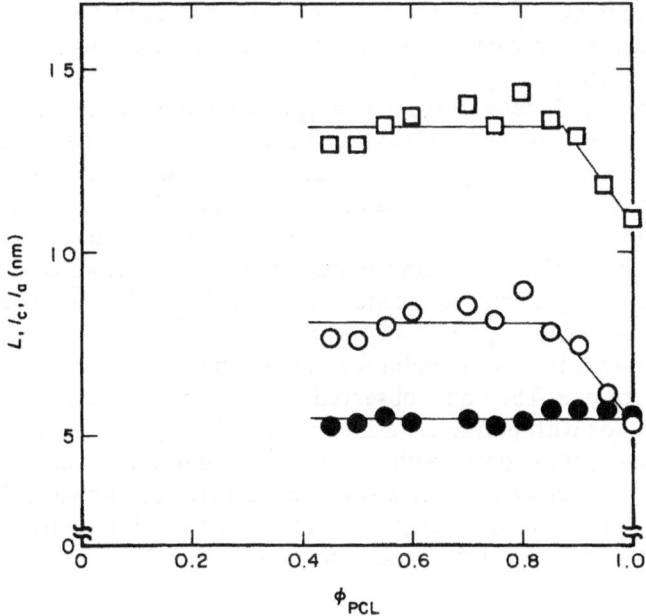

Fig. 63. Dependencies of long period, L (□), lamellar thickness, l_c (●) and amorphous layer thickness, l_a (○) with PCL content in blends of PCL with oligo-PS; taken from [145]

crossed the binodal and the melt phase-separated by nucleation and growth to create the surface droplets. Ultimately, the droplets formed a pure PS phase.

Li et al. [81] also studied the intermediate regime B in which samples were cooled into the metastable region of the phase diagram. Liquid-liquid phase separation gave spherical droplets of a PS minor phase in a PCL-rich matrix. The number of droplets increased and their size decreased with increased degree of undercooling; the PS droplets were the minor phase. The PCL then crystallised from the major phase, excluding PS. The PS droplets became engulfed by the growing spherulites and some were deformed in the process; spherulites formed were irregular. At lower crystallisation temperatures the glassy PS droplets were not deformed and were contained within the spherulites. At higher crystallisation temperatures small PS droplets formed, from the PCL-rich matrix, at the surfaces of the spherulites, as in regime C.

Both Tanaka and Nishi [22, 146] (30 wt % PS) and Li et al. [81] (20 wt % PS) studied the kinetics of crystallisation in these blends. Both groups found constant rates of radial growth at larger degrees of undercooling (35–47 °C) but at smaller degrees of undercooling they both found that rates of radial growth decreased with crystallisation time. The non-linear kinetics correlated with the development of PS droplets at the growth surfaces of the spherulites. While observations by the two groups were similar, their interpretations were different. Thus, Tanaka and Nishi analysed their data in terms of reduced rates of crystallisation due to increased PS content in the homogeneous material at the growth surface. Li et al. interpreted their data in terms of additional work done in moving the PS droplets at the growth surface.

Nojima et al. [145] determined the long period in the crystalline regions from scattering data and separated it into the thicknesses of lamellae and amorphous layers in the crystalline regions. The lamellar thickness was independent of blend composition but the thickness of the amorphous layer increased at low oligo-PS contents and then became constant (Fig. 63). Nojima et al. [145] discussed the composition of the amorphous layers under various conditions.

Watanabe et al. [82] studied blends of PCL ($\overline{M}_n$=10,700, $\overline{M}_w$=33,000) with polystyrene ($\overline{M}_n$=460, $\overline{M}_w$=550; $\overline{M}_n$=840, $\overline{M}_w$=950; $\overline{M}_n$=2,700, $\overline{M}_w$=2,800) and found depressions in melting point for PCL for the low molecular weight samples of polystyrene. They also observed UCST behaviour and found critical points for blends with polystyrene at about 95 °C ($\overline{M}_w$=550) and about 230 °C ($\overline{M}_w$=950), at ϕ_{PS} about 0.9 in both cases; for the highest molecular weight polystyrene only a part of the binodal could be detected between 200 °C and 300 °C at low polystyrene contents. From these data an interaction parameter of χ=(−0.469/32)(1−2.46×10^4/T) was determined (Sect. 7).

17.2
Blends with Poly(styrene-*co*-allyl-alcohol)

Copolymers of styrene and allyl alcohol have the general structure **18** with the two different units in different proportions, x and y.

Polystyrene is generally immiscible with PCL, except for oligomers. One method of enhancing miscibility is to introduce functional groups capable of interacting with the carbonyl group of PCL. Hydroxyl groups are one possibility, in a similar manner to that used in the case of phenoxy resin. This method has been adopted in the study of blends of PCL with copolymers of styrene and allyl alcohol (SAA) **18**. Barnum et al. [92] used such copolymers with hydroxyl contents of 1.3–7.7 wt %. The polymers had low molecular weights (Table 16); indeed they were little larger than the polystyrene oligomers studied by Nojima et al. [145]. The glass transition temperatures of SAA samples varied with both hydroxyl content and with molecular weight in this molecular weight regime.

Blends of PCL-700 (Table 1) with the SAA copolymers were prepared by melt mixing, stirring by hand in view of the low T_gs, at about 110 °C and then cooled.

Structure 18

Table 16. Characterisitics of styrene-allyl alcohol copolymers blended with PCL; data taken from [92]

Hydroxyl content (wt %)	7.7	5.7	2.5	1.3
$10^{-3}\,\overline{M}_w$	1.70	2.34	1.42	2.10
$10^{-3}\,\overline{M}_n$	1.15	1.60	0.95	1.40

Glass transitions were determined by DTA; enthalpies of fusion and melting points of PCL were determined by DSC. Crystallisation was allowed to proceed at 30 °C for 30 min prior to determining melting points. Each blend had a single, composition-dependent glass-transition temperature. Values of T_g fell just below the straight line joining the T_gs of the components, indicating only weak interactions between the components [92].

For blends with 50 wt % or more PCL, enthalpies of fusion and melting points were determined; enthalpies of fusion were proportional to the weight fraction of PCL. From the enthalpies of fusion and melting point depressions (without correction using Hoffman-Weeks plots to determine equilibrium melting temperatures) estimates were made of the interaction parameter B (Eq. 8). Values of B were found to be negative indicating a negative enthalpy of mixing. The variation of B with hydroxyl content had a minimum at about 3 wt % hydroxyl and curves indicated positive values of B, and positive enthalpies of mixing, for hydroxyl contents greater than 10 wt %. The results suggest that favourable interactions promoting miscibility with PCL only exist for a limited range of hydroxyl contents.

17.3
Blends with Poly(vinyl phenol) or Poly(4-hydroxystyrene) (P4HS) and its Styrene Copolymers

A second approach to enhancing miscibility with styrene-type polymers is to introduce H-bonding groups directly into the polystyrene. This approach has been adopted by Moskala et al. [89] who stated that PCL is miscible with poly(vinyl phenol) which has a hydroxyl group in the aromatic group of polystyrene, i.e. is poly(4-hydroxystyrene) (P4HS) **19**.

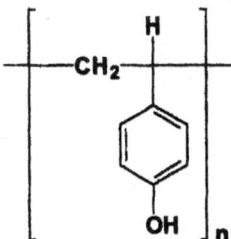

Structure 19

These workers investigated the infrared spectra of the blends in order to obtain evidence for interactions (Sect. 7). Amorphous PCL (at 75 °C) showed a carbonyl stretching band at 1734 cm^{-1}. Addition of P4HS caused the appearance of a new band at 1708 cm^{-1} which increased in intensity with increasing P4HS content and became dominant in blends containing more than 70 wt % P4HS. This band was attributed to PCL carbonyl groups H-bonded to P4HS. At room temperature, where PCL is partially crystalline, the IR spectrum of pure PCL shows a carbonyl stretching band at 1724 cm^{-1} from crystalline PCL. Increased contents of P4HS at room temperature again caused the appearance of an absorption band at 1708 cm^{-1} at the expense of amorphous PCL, primarily at low P4HS contents. At high P4HS contents, greater than about 50 wt % P4HS, the intensity of the band for crystalline PCL decreased in intensity and was small at 10 wt % PCL. These workers also investigated bands corresponding to hydroxyl stretching. They interpreted the spectrum for pure P4HS as showing a band at 3525 cm^{-1}, due to free hydroxyl groups, and a broad band centred at 3370 cm^{-1}, due to a variety of H-bonded hydroxyls in self-association. In both amorphous and partially crystalline blends with PCL the main change in hydroxyl stretching absorptions was a shift in the position of the maximum of the broad peak to lower frequencies, indicating a change in the nature of the H-bonded associations.

Lezcano et al. investigated the melting behaviour of PCL ($\overline{M}_n$=50,000, $\overline{M}_w/\overline{M}_n$=1.6) in its blends with P4HS ($\overline{M}_n$=1,500, $\overline{M}_w/\overline{M}_n$=2) [42]. Samples rich in P4HS (<30 wt % PCL) showed a single composition-dependent glass-transition temperature, while samples rich in PCL showed both a T_g and crystal melting endotherm by DSC; samples were initially cooled after heating to 150° C for 10 min. Samples quenched from the melt exhibited a single composition-dependent T_g. Values of T_m were depressed relative to that of pure PCL. After using Hoffman-Weeks plots to determine equilibrium melting points and analysing the data by the Nishi-Wang equation (Eq. 9), Lezcano et al. obtained an estimate of the binary interaction parameter which was negative (consistent with polymer-polymer miscibility) [42]. Values of interaction parameters are given in Sect. 7.

Because PCL is immiscible with polystyrene and miscible with P4HS it is logical to examine the miscibility of PCL with copolymers of styrene and 4-hydroxystyrene 20.

Structure 20

A study of this type was undertaken by Vaidya et al. [147]. These workers prepared P4HS and copolymers with styrene (P4HS-x where x defines the mole per cent of 4HS in the copolymer); values of x were 1–9. The copolymers all had number-average molecular weights between 11,000 and 25,000 g mol^{-1} ($\overline{M}_w/\overline{M}_n$=1.7); glass-transition temperatures of the copolymers varied from 101 °C for P4HS-1 to 114 °C for P4HS-9. Blends were cast from solutions in THF and dried under vacuum at 35 °C for 5 days. DSC measurements were undertaken to determine T_gs and equilibrium melting points for PCL ($\overline{M}_w$=33,000) in the several blends. These data demonstrated that P4HS-1 copolymers are miscible with PCL only at high P4HS-1 (>78 wt %) contents; phase separation occurred above about 65 °C. With increasing 4HS contents the miscibility increased and blends with more than 60 wt % P4HS-5 were miscible; phase separation occurred above about 75 °C at 60 wt % P4HS-5 and at about 210 °C at 95 wt % P4HS-5. Samples containing more than 7 mol % 4HS were miscible across the whole composition range up to temperatures of about 190 °C for P4HS-7 and higher for P4HS-9. Values of the composition-dependent T_gs of the miscible blends showed some deviation from Eq. (23) which was attributed to the interactions associated with the presence of hydrogen bonding. Phase diagrams summarising the data and obtained from DSC and cloud point data were presented. Thus, these various samples were characterised by LCST behaviour, with the LCST located below room temperature for blends with P4HS with the lowest 4HS contents. It was also found that PCL crystallised from the miscible blends with P4HS-7 and P4HS-9 under certain conditions, especially at very high PCL contents – less than 50 wt % P4HS-7, 40 wt % P4HS-9 and 20 wt % P4HS-100. Degrees of crystallinity of the PCL were low. These workers also calculated phase diagrams for the different blends and these were generally in accord with experimental observations [147].

17.4
Blends with Methoxystyrene-Hydroxystyrene Copolymer

Sanchis et al. [91] investigated interactions and miscibility of partially methoxylated P4HS (PMHS) with PCL ($\overline{M}_n$=50,000, $\overline{M}_w/\overline{M}_n$=1.6). The PMHS was prepared by methoxylation of P4HS ($\overline{M}_n$=1,500, $\overline{M}_w/\overline{M}_n$=2) to 60% conversion of the hydroxy groups (21 with y=0.6, x=0.4). Thus the PMHS retained the ability of P4HS to hydrogen-bond with PCL but to a reduced extent and the potential extent of intra- and inter-molecular hydrogen-bonding changed. Blends were prepared by casting from solution in THF and after drying for several days at room temperature the samples were dried under vacuum at 60 °C for 10 days.

DSC determinations were made on samples heated to remove crystallinity and cooled to 30 °C to crystallise the PCL. PMHS exhibited a T_g at 100 °C. Blends with 60 wt % or more PCL were partially crystalline and all samples had a single T_g, indicating complete miscibility in the amorphous phase. Crystallinities of PCL in samples containing 70 wt % or more PCL were about 50%, similar to that in pure PCL. The T_gs, after correction for the presence of crystalline PCL,

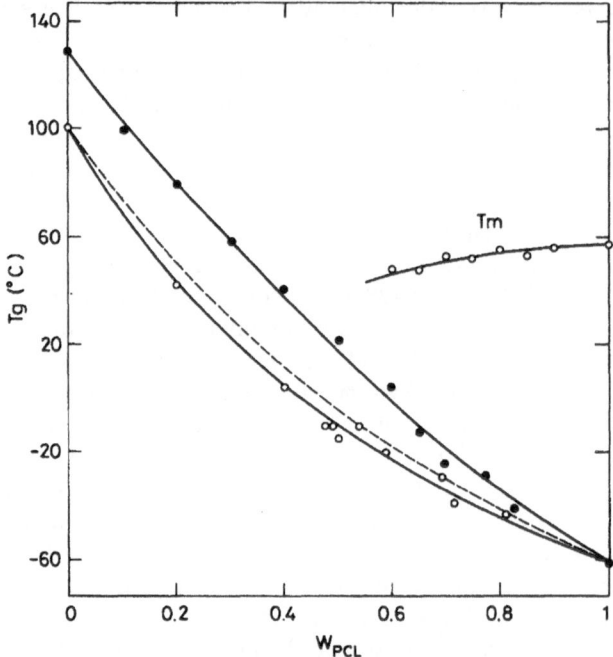

Fig. 64. Dependence of melting temperature (T_m) and glass-transition temperature of PCL/PMHS blends (○) with weight fraction of PCL, crystallised at 30° C; *dashed line* is calculated from Eq. (23). Also included is the variation in glass-transition temperature for PCL/P4HS blends; data taken from [130]; figure taken from [91]

Structure 21

showed small negative deviations from Eq. (23). Figure 64 summarises the T_g and T_m data for blends of PCL with P4HS and PMHS; data for P4HS blends were taken from [42]. This negative deviation was attributed to a balance between two factors. Formation of strong hydrogen-bonding between different molecular species would reduce free volume and molecular mobility and would tend to produce a positive deviation in T_g while the destruction of self-association hydrogen-bonding would enhance free volume and produce a negative deviation. In contrast with P4HS blends, blends of PMHS with PCL retained crystallinity to lower PCL contents (60 wt %) than P4HS blends (75 wt % PCL). From values of melting point depression, interpreted in terms of the Nishi-Wang equation (Eq. 10), values of interaction parameters were determined. A negative value for the polymer-polymer interaction parameter was determined with values quoted in Sect. 7. The value of χ determined was consistent with miscibility in the melt but was not as low as the value found for poly(4-hydroxystyrene) blends with PCL and this is consistent with lower content of groups capable of forming hydrogen-bonds. Thus, the presence of methoxy groups increases the interaction parameter and opposes miscibility.

Interactions between PMHS and PCL were determined by FTIR. In the carbonyl stretching region PCL showed an absorption at 1736 cm^{-1} for amorphous PCL and, at lower temperatures, an absorption for crystalline polymer at 1724 cm^{-1}. In the blends, an additional band at 1708 cm^{-1} was assigned to the hydrogen-bonding of the carbonyl group to hydroxy groups in P4HS or PMHS. In PMHS/PCL 40/60, as well as in P4HS/PCL 20/80 blends, there was an additional absorption at 1708 cm^{-1} assigned to the hydrogen-bonded carbonyl group, confirming the existence of interactions between components. The intensity of the band at 1708 cm^{-1} increased with increasing P4HS or PMHS content and at higher concentrations of P4HS or PMHS the crystalline band at 1724 cm^{-1} was absent; results were in accord with previous observations of Moskala et al. [89] for P4HS/PCL blends; however these workers did observe evidence of PCL crystallinity in blends containing as little as 10 wt % PCL, presumably due to different conditions for sample preparation. Quantitative assessment of the interactions indicated that a smaller proportion of PCL carbonyl groups were hydrogen-bonded in PMHS/PCL blends than in P4HS/PCL blends, consistent with the lower concentration of interacting groups in PMHS. From the frequencies of the carbonyl stretching, the strengths of hydrogen-bonds in the two systems appears to be the same.

As with P4HS, PMHS has a broad absorption band (3100–3500 cm^{-1}) due to self-associating hydroxyl groups. In addition, a shoulder at 3525 cm^{-1} was attributed to free (non-hydrogen-bonded) hydroxyl groups. In the PMHS/PCL blends the band at 3525 cm^{-1} disappeared and the broad band shifted to higher frequencies and narrowed due to hydroxyl groups becoming hydrogen-bonded to PCL carbonyl and similar behaviour was observed for P4HS/PCL blends. In the blends, it was concluded, all hydroxyl groups were hydrogen-bonded and interactions with PCL were stronger than self-associations. The destruction of self-association enhances the entropy of mixing and favours miscibility as does the negative enthalpy of mixing which presumably accompanies the development of strong interactions between the components.

At higher temperatures (70 °C), in the molten state, a dominant band for hydroxyl stretching was attributed to hydroxyl hydrogen-bonded to carbonyl and a weak band to hydroxyl hydrogen-bonded to hydroxyl. The carbonyl stretching region showed a lack of PCL crystallinity and carbonyl groups were primarily hydrogen-bonded to hydroxyl, but there was a reduction in the proportion of carbonyl groups hydrogen-bonded to hydroxyl and interactions were destroyed at the higher temperature where thermal energy is greater [91]. Observations were in accord with those from inverse-phase gas chromatography [90].

17.5
Blends with Styrene-Maleic Anhydride Copolymer

Studies have also been performed on copolymers of styrene with maleic anhydride (SMA-n) 22 containing n wt % maleic anhydride.

Defieuw et al. [148] investigated blends of PCL ($\overline{M}_n$=14,000, $\overline{M}_w/\overline{M}_n$=1.6) with two styrene-maleic anhydride copolymers containing SMA-14 ($\overline{M}_n$= 59,000, $\overline{M}_w/\overline{M}_n$=2.26) and SMA-25 ($\overline{M}_n$=50,000, $\overline{M}_w/\overline{M}_n$=2.35). The glass-transition temperature of SMA-14 was about 140 °C and of SMA-25 about 160 °C, as determined by dynamic mechanical analysis.

Blends containing 30–90 wt % of the SMA copolymers were prepared by co-precipitation from homogeneous solution. When quenched from elevated temperature, the blends gave optically clear samples through the temperature range 60–200 °C. These samples gave single, composition-dependent glass-transition temperatures when studied by dynamic mechanical analysis. Thus, the components were found to be miscible in the melt and could be retained in the homogeneous state when quenched [148].

Vanneste and Groeninckx investigated the miscibility of SMA copolymers with various MA contents with PCL. These workers found SMA-2 to be immiscible with PCL. SMA-17 and SMA-28 were found to be totally miscible with PCL while SMA-8 and SMA-14 exhibited partial miscibility and LCST behaviour [149]. The discrepancy with the previous work [148] was stated to be because these latter studies used polymers of higher (unspecified) molecular weight [149].

SMA copolymers appear to nucleate crystallisation of PCL from the melt. For blends containing >20 wt % SMA copolymer the high nucleation density prevented the determination of spherulite growth rates and thin films could not be studied by low-angle-laser-light-scattering because of orientation effects [148].

For blends based on SMA, during isothermal crystallisation at 45 °C, rates of radial growth of spherulites were constant. Spherulite radial growth rates increased with increasing PCL content and, for blends with equal SMA content (10 wt %), that based on SMA-14 showed higher rates of radial growth than that based on SMA-25 (Fig. 65). This latter effect was attributed to greater segmental mobility in the SMA copolymer having lower maleic anhydride content [148].

Structure 22

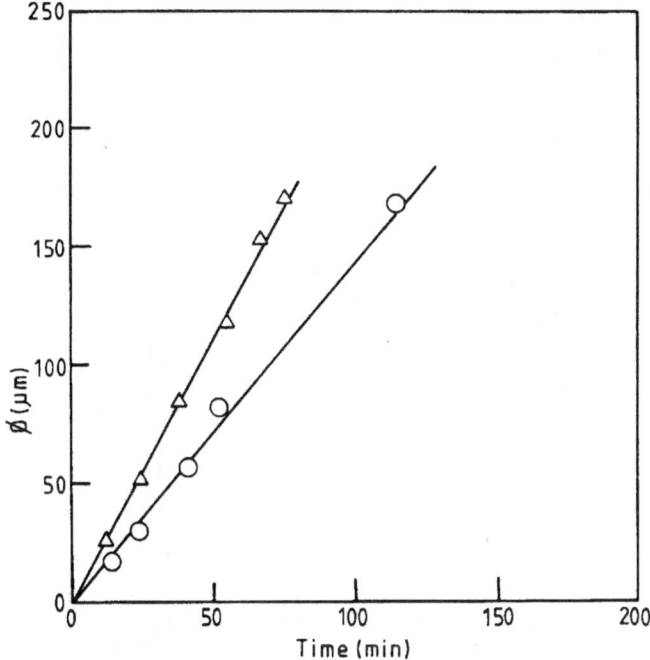

Fig. 65. Development of spherulite size φ with time for PCL crystallisation in blends with SMA with 90 wt % PCL, (△) SMA-14 and (○) SMA-25; taken from [148]

The crystalline blends consisted of space-filling spherulites with a long-spacing of 130–150 Å, approximately. Optical micrographs showed no evidence of segregation of SMA copolymer into discrete regions (interspherulitic) and this observation, together with the observation of constant rates of radial growth, suggested that SMA segregated between lamellae and fibrils as the spherulites developed. This view was supported by the decrease, with increasing SMA content, of the corrected values of angles of maximum scattering intensity, determined by small-angle X-ray scattering; the long-spacing of the lamellae increased with SMA content. Further support for this view was obtained from one-dimensional correlation functions. The lack of large-scale segregation of SMA (cf. blends with phenoxy (Sect. 16), for example) was attributed to the inherent low segmental mobility in SMA copolymers; the effect could also have arisen from a large value of the polymer-polymer interaction parameter but no relevant data are available [148].

Double melting peaks were observed for blends with ≥50 wt % PCL in blends with SMA-14 and ≥60 wt % PCL in blends with SMA-25; the lower-temperature peak was the smaller peak. Values of crystal-melting temperature (the major peak at higher temperature) increased with increasing PCL content from about 58 °C to 61.5 °C. (Fig. 66). It was reported that this effect might have arisen from thermodynamic melting point depression or from variations in lamellar thick-

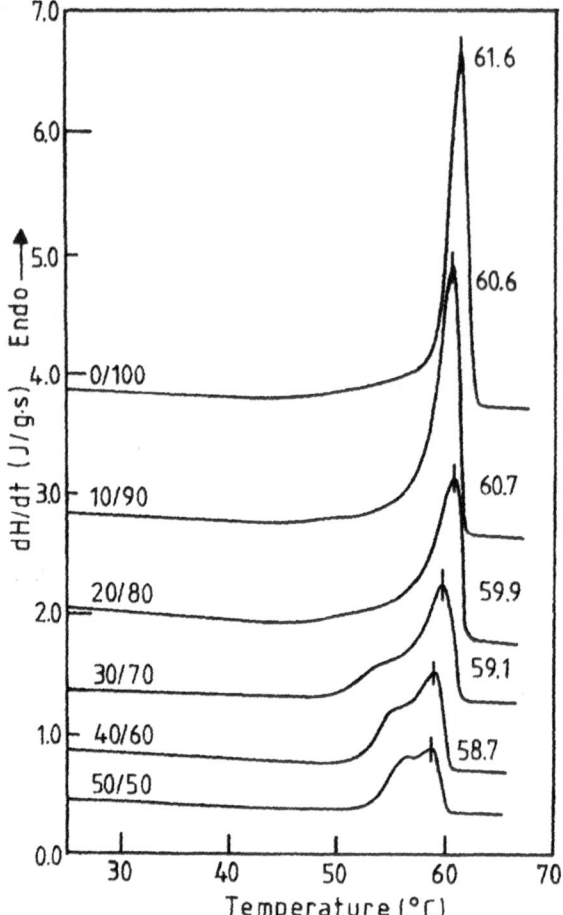

Fig. 66. DSC thermograms for PCL/SMA-14 blends after crystallisation at 35 °C for 14 days; taken from [148]

ness at lower PCL contents. The lower-temperature, crystal-melting peak was not observed immediately after crystallisation (for, say, 30 min) but developed on more prolonged crystallisation. This peak, which increased in magnitude and shifted to higher temperatures with increasing crystallisation times, up to, say, 1500 minutes, was attributed to secondary crystallisation from higher SMA-content mixtures and the formation of less perfect crystallites [148].

Crystal melting enthalpies were also determined and they decreased with increasing SMA content and became zero at 50–60 wt % SMA. (Fig. 67) The more rapid loss of crystallinity in SMA-25 blends was attributed (again) to the higher glass-transition temperature of SAM-25, which reduced the level of crystallinity; molecular mobility was reduced at higher PCL contents in the amorphous phase.

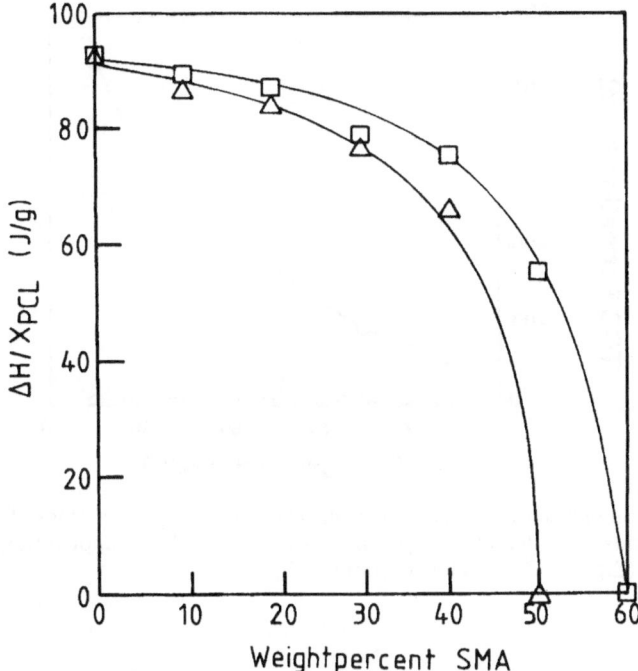

Fig. 67. Variations in enthalpy of melting of PCL in PCL/SMA blends, crystallised as in Fig. 56. (□) SMA-14 and (△) SMA-25; taken from [148]

18
Blends with Other Polymers

There have been several smaller studies of blends of PCL with a variety of other polymers. These studies have often considered specific features of the blends and are not as comprehensive as many of the studies already considered.

18.1
Blends with Polyolefines

Polyolefines are major-tonnage polymers and are immiscible with PCL. Nevertheless their blends with PCL have attracted attention because of the biodegradability of PCL. The PCL, of course, does not introduce biodegradability into the polyolefines but its degradation can render the polyolefine blends disintegrable. Because polyolefines enter the environment through packaging waste their disintegration can reduce the undesirable visual impact of that waste.

Thus, Clendening and Potts [150] claimed disintegrability of blends of PCL with low-density polyethylene and with polypropylene. Subsequently, Iwamoto and Tokiwa [151] studied blends of PCL with low-density polyethylene and ob-

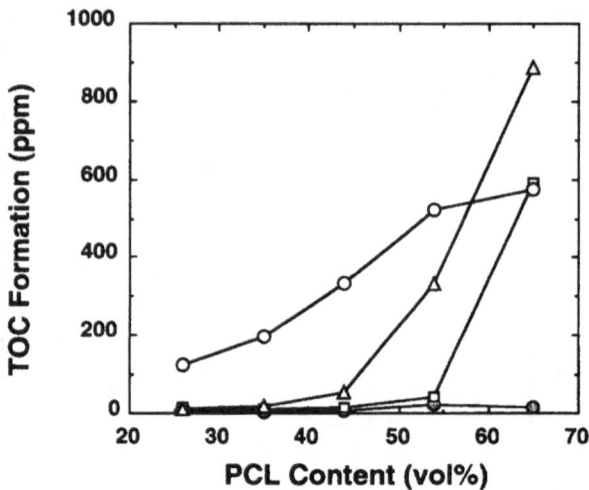

Fig. 68. Variation in water soluble total organic carbon (TOC) as functions of PCL content in PCL/PP blends for different ratios of melt viscosities of the components: η_{PP}/η_{PCL} (●) 0.20, (□) 0.67, (△) 1.0, (○) 0.33; taken from [152]

served that disintegrability of the blends depended on their composition and the melt viscosities of the components. Iwamoto and Tokiwa also investigated blends of PCL with polypropylene. In the former study, these authors inferred that the disintegrability of the blends depended on the phase structure and required the PCL to be the continuous (matrix) phase. This aspect was investigated directly in the latter study when samples of different compositions were prepared from combinations of polymers with four different ratios of melt viscosity at a blending temperature of 230 °C; samples were prepared at a constant shear rate of 60 s^{-1} in an extruder.

Mechanical blending of two immiscible polymers might be expected to produce a dispersion of the minor component, as spherical domains, within a matrix of the major component; the size of the dispersed phase might be expected to depend on the amount of mechanical work expended in blend preparation. However, as we have already noted (Sect. 5) rheological factors, in particular the relative viscosities of the components, are also important in determining phase morphology. Iwamoto and Tokiwa [152] applied the relationship at Eq. (20) to a consideration of PCL/polypropylene blends and their susceptibility to enzymatic degradation. Optical microscopy showed that sample morphologies were as predicted by values of α, except for $\phi_i > 0.75$, when the major phase was always the continuous phase, as predicted. Enzymatic degradability of the blends was estimated from the yields of water-soluble organic carbon developed when treated with *R. arhizus* lipase in the presence of a surfactant at 35 °C. It was demonstrated that the development of water-soluble organic carbon increased with PCL content in some but not all blends (Fig. 68). However, they observed a strong dependence of soluble carbon formation with α. Essentially no degradation occurred for α<0.8 but degradation increased rapidly with α for α>1. The combined results of this study established that enzymatic degradation only occurred in these blends when PCL was the (or a) continuous phase.

Blends of PCL with linear low-density polyethylene have been prepared and their rheological properties were investigated. Data demonstrated that in variations of ln(viscosity) with composition of the blends experimental values fell below the weighted average values. A model was proposed to describe this negative deviation [153].

18.2
Blends with Poly(vinyl methyl ether) (PVME)

Blends of PCL with PVME 23 have been investigated in conjunction with studies of ternary blends (Sect. 20).

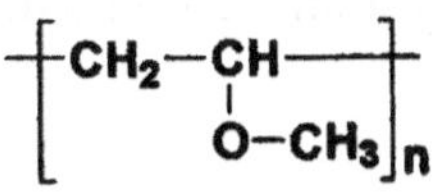

$$\left[CH_2-\underset{\underset{O-CH_3}{|}}{CH} \right]_n$$

Structure 23

Watanabe et al. [82] demonstrated that T_m for PCL is depressed in mixtures with PVME up to 80 wt % PVME. Above T_m they found that PCL ($\overline{M}_n$=10,700, $\overline{M}_w$=33,000) is miscible with PVME ($\overline{M}_n$=46,500, $\overline{M}_w$=99,000) to a critical point at about 200 °C above which the system exhibits UCST behaviour. From the phase diagram they obtained a value of the interaction parameter as $\chi=$ (−9.32/1.8)(1−396.5/T). The binodal calculated from this value agreed well with cloud point data; spinodals were also calculated.

Quipeng [169] demonstrated that glass-transition temperatures for PCL/PVME mixtures were single and composition dependent; these authors made no reference to crystal melting temperatures. They observed LCST behaviour for this system with a critical point at about 200 °C for a 50:50 w/w mixture. This study was part of an investigation of ternary polymer blends (Sect. 20).

18.3
Blends with Poly(N-phenyl-2-hydroxytrimethyleneamine)

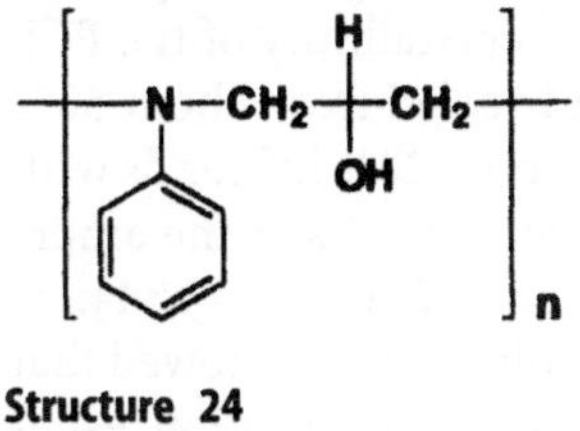

$$\left[-\underset{\underset{\bigcirc}{|}}{N}-CH_2-\underset{\underset{OH}{|}}{\overset{\overset{H}{|}}{C}}H_2 \right]_n$$

Structure 24

Guo et al. [154] synthesised poly(*N*-phenyl-2-hydroxytrimethyleneamine) **24** ($\overline{M}_w$=19,600, $\overline{M}_w/\overline{M}_n$=1.37) from epichlorohydrin and aniline. This polymer, amorphous with a glass-transition temperature of 70 °C, was blended with PCL ($\overline{M}_n$=70,000–100,000) by solution casting from chloroform; samples were dried under vacuum at 50 °C for 2 weeks. The pendant hydroxyl group in **24** is capable of forming interactions with other polymers. FTIR studies of the polymers showed that the hydrogen-bonded hydroxyl stretch at 3304 cm^{-1} of **24** shifted to higher frequency in the presence of PCL and the carbonyl stretching band at 1725 cm^{-1} in pure PCL was partly replaced by a band at 1710 cm^{-1} in the blends; these data were to taken to indicate hydrogen bonding between the pendant hydroxyl groups of **24** and the carbonyl of PCL. Blends exhibited a single composition-dependent T_g; T_gs were below room temperature for blends containing more than about 25 wt % PCL. Samples containing more than 60 wt % PCL crystallised at or above room temperature (DSC endotherms). PCL crystallinities in the blend were estimated at about 60% in blends containing up to 20 wt % **24** but only about 3% in blends containing more **24**. After allowing for the presence of crystalline PCL, measured T_gs for amorphous material corresponded well with Eq. (23).

18.4
Blends with Phenylacetylene-Carbon Monoxide Alternating Copolymer

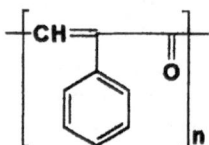

Structure 25

Chen et al. [155] synthesised a copolymer of phenylacetylene and carbon monoxide ($\overline{M}_n$=197,300, $\overline{M}_w$=414,200, T_g approximately 110 °C) and blended this with PCL ($\overline{M}_n$=34,000, $\overline{M}_w$=6,900) by coprecipitation from toluene solution into methanol. The blends were washed and dried under vacuum for 2 days at 40 °C. Blends were annealed at 115 °C for 3 min, followed by quenching, to erase thermal history. Samples were than examined by DSC with a heating rate of 20 °C min^{-1}. The blends exhibited a single composition-dependent T_g, indicating miscibility over the whole composition range. In blends containing more than 60 wt % PCL some PCL crystallised. The degree of crystallinity of the PCL in the blends was determined from heats of fusion and varied from about 50% in blends containing more than 60 wt % PCL, falling to about 20% in blends with 50 wt % PCL. After correction for the crystalline PCL content, T_gs of the amorphous component showed reasonable agreement with Eq. (22) with *k*=0.344.

Peak temperatures of crystallisation peaks in DSC cooling curves showed that the presence of the copolymer depressed the onset of PCL crystallisation, especially when the copolymer content exceeded 20 wt %. These data were consistent with reduced rates of crystallisation in the presence of a miscible component of higher T_g which served to reduce molecular mobility and the rate of crystallisation. The blends exhibited two melting endotherms, the lower being attributed to the melting of ill-formed crystalline material and the higher to the melting of recrystallised material [155].

18.5
Blends with Liquid-Crystalline Polymers

Thermotropic, main-chain, liquid-crystalline polymers (LCPs) have attracted considerable attention as a result of their high stiffness and mechanical properties. There has been interest in combining the LCPs with other materials. In one area, LCPs are used, in relatively low concentration, to reinforce less-stiff materials. In another case, a second component is used as a solvent to increase the mobility in the LCP and form lyotropic liquid-crystalline materials. There are now two reports, from Kricheldorf's group, of blends of PCL with liquid-crystalline polyesters [156, 157].

In the first report, by Taesler et al. [156], a copolyester **26** was added (up to 10 wt %) of PCL. Alone **26** forms a liquid-crystalline melt at 207 °C and becomes isotropic (with decomposition) at 440 °C.

Blends with PCL-700 (Table 1) were formed by coprecipitation from solutions in dichloromethane/trifluoroacetic acid (7:1) into methanol, washed with methanol and dried under vacuum for 12 h. To provide samples for mechanical testing, samples were hot-pressed at 70 °C and quenched at 0 °C. The blends (with 1–10 wt % PCL) exhibited melting endotherms, corresponding to PCL melting at 60–65 °C; crystallinities fell below those expected for two pure phases. Optical microscopy of samples at temperatures above the melting point of PCL and up to 200 °C showed birefringent areas, which oriented under shear, typical of liquid crystallinity, in isolated regions. On cooling samples with 1 wt % LCP from 70 °C the PCL crystallised as spherulitic domains outside the birefringent areas. On cooling from higher temperatures liquid-crystalline regions persisted within the PCL spherulites. These results indicated a pure PCL phase and a dispersed

Structure 26

lyotropic liquid-crystalline phase in which PCL acted as solvent. Samples cooled from the melt showed the dispersed phase trapped within the PCL spherulites. It was concluded that the LCP-phase was a mixed PCL/LCP phase because it was possible to induce orientation at lower temperatures and with less shear than required for the pure LCP phase. Above 200 °C the domains developed a more fibrillar texture, typical of the pure liquid-crystalline polyester. Thus, it was considered that above 200 °C the LCP separated as a pure phase.

Addition of only 1 wt % LCP to PCL increased Young's modulus by a factor of two and reduced the extension to break from 600% to about 7%. The effect at high concentrations of LCP was proportionally less and 10 wt % LCP increased Young's modulus by a factor of four and reduced elongation to break to about 2% [156].

Subsequently, Kricheldorf et al. [157] prepared a series of copolyesters and blended them with PCL using the same coprecipitation procedure. In conventional stress-strain measurements, polyesters 27a–c (4 wt %) all raised the initial modulus of the PCL but 27c embrittled the blend which did not yield but fractured at a strain of 0.04. The polyester27a, which forms thermotropic liquid crystalline phase at high temperature, at 4 wt % raised the tensile storage modulus of PCL by 80% at low temperatures (–100 °C) and similar results were obtained with 26. Polyester 27b, which did not itself form a liquid crystalline phase, also reinforced PCL. On a weight basis the reinforcing effect of 27b as less than that of 27a but on a molar basis the effects were the same. Thus, the rigid-rod polyesters were equally effective in molar terms despite one of them not showing liquid crystalline characteristics itself. It was shown that polyesters with long aliphatic side chains 28 were ineffective in raising the elastic modulus whereas those polymers with pendant aromatic units, especially p-substituted, or with rigid biphenyl units 26 were most effective in raising the storage modulus.

Rheological measurements showed that the viscoelastic behaviour of a blend containing 1 wt % 28b was similar to that of pure PCL at high frequencies, but at lower frequencies ($\omega a_T < 2$ rad s^{-1}) the storage shear modulus was much higher than that of PCL (Fig. 69). No interpretation of the detailed rheological behaviour of the blends was offered.

Structure 27

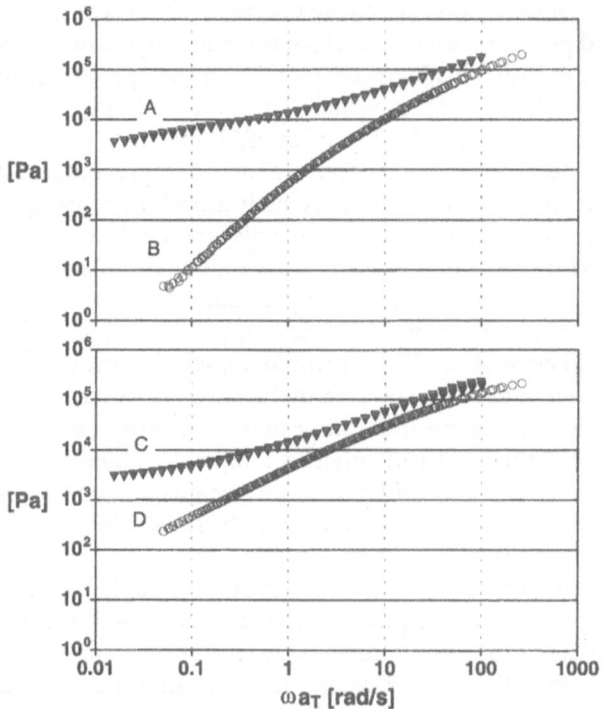

Fig. 69. Results of rheological measurements (80° C) on PCL and its blend with 1 wt % of 28b: curve A – storage modulus of blend; curve B – storage modulus of PCL; curve C – loss modulus of blend; curve D – loss modulus of PCL; taken from [157]

Structure 28

Optical microscopy showed that even 1 wt % of rigid-rod polyester could generate birefringence in 15–25 vol.% of the sample, which became a mobile anisotropic phase. A blend with 4 wt % of **27a** induced birefringence throughout the blend. The observed birefringence was diffuse and not associated with a typical texture of a liquid crystalline phase. This anisotropic phase was stable at 150 °C and only vanished as the sample temperature approached T_m of the rigid-rod polymer. These observations are consistent with phase separation, at about T_m of the rigid-rod polymer, in a blend which was a homogeneous dispersion of rigid-rod polymer in PCL. This is consistent with the accepted immiscibility of liquid crystalline and isotropic materials. Thus, a mobile birefringent phase appeared to be stable between the T_ms of the PCL matrix polymer and the rigid-rod polymer. These observations were made on coprecipitated material and were not observed when polymers were melt mixed. In coprecipitated samples the materials existed as almost molecular dispersions of rigid-rod polymer in PCL, possibly with small bundles of rigid-rod polymer molecules aggregating to give 'ordered stacks'. What is particularly notable is that optical anisotropy was observed in systems containing the rigid-rod polymers 27c, 28b which are not inherently liquid crystalline. Thus, it would appear that the small assemblies of rigid-rod polymer chains are solvated by PCL chains which align with them to induce anisotropy. In this way the volume of the anisotropic phase can be greater than the volume of the rigid-rod polymer component alone. This association in an anisotropic phase was considered to be responsible for the significant improvement in mechanical properties achieved by the addition of only 1 wt % rigid-rod polymer to PCL.

It was also remarked that the formation of a lyotropic blend from PCL and 27c, two isotropic components, was particularly notable. It was suggested that, as formation of a liquid-crystalline phase is strongly dependent on free-volume fraction, the combination of PCL chains aligned with rigid-rod polymer chains can reduce the overall free volume, induce nematic order and stiffen the aggregate [157].

19
Blends with Block Copolymers

19.1
Blends with Sulfonated Styrene-(Ethylene-co-Butylene)-Styrene Copolymer

Block copolymers with pendant acid groups, which can be converted to their metal salts, have been prepared. One such polymer was the styrene-(ethylene-co-butylene)-styrene triblock copolymer in which 8.3% of the styrene units in the terminal styrene blocks had been sulfonated 29. This polymer ($\overline{M}_n$=50,000, 29.8 wt % styrene, initially) (SSEBS) was blended with PCL ($\overline{M}_n$=40,000) [158].

Structure 29

The SSEBS copolymer alone exhibited a single broad peak when investigated by SAXS. Addition of PCL brought about the appearance of several scattering peaks which increased in intensity and definition to about 25 wt % PCL. This scattering pattern was consistent with, and indicated the formation of, a well-defined lamellar microstructure in the blend. The scattering peaks shifted to lower angles with increasing PCL content, consistent with swelling of a microphase in the lamellar structure. Blends with 15 wt % and 27 wt % PCL were reported to be transparent. With 59 wt % PCL the microstructure lost order; evidence for retention of a lamellar structure persisted but the scattering peaks were broadened. At this higher PCL content, evidence for crystalline PCL in the system was obtained; a double crystal melting endotherm was observed at about 50 °C. Some evidence for a little PCL crystallinity in samples with 27 wt % PCL was obtained.

DSC data for SSEBS gave a large broad endotherm at about 0 °C which was unattributed. T_g for the SSEBS hard block (sulphonated polystyrene) was at about 64 °C and decreased to 54 °C on addition of 27 wt % PCL to the copolymer; T_g of the ethylene-butylene block was about −40 °C. This change in T_g of the hard block was attributed to disruption of ionic interactions in the sulphonated polystyrene microphase. It was also suggested that swelling of the sulphonated polystyrene blocks by PCL also allowed more facile development of a well-ordered lamellar morphology.

19.2
Blends with Caprolactone Block Copolymers

There has been considerable general interest in blends of block copolymers with one homopolymer which is chemically identical with one block of the copolymer. Two such systems have been reported for blends of PCL with block copolymers in which one block is PCL.

19.2.1
ε-Caprolactone-2,2-Dimethyltrimethylene Carbonate Diblock Copolymers

Kummerlöwe and Kammer reported observations on the form of the crystalline structure in blends of PCL ($\overline{M}_n$=40,000) with a diblock copolymer of PCL ($\overline{M}_n$= 20,000) and poly(2,2-dimethyltrimethylene carbonate) (PDTC) 30 ($\overline{M}_n$=20,000); both components of the block copolymer are crystallisable.

Structure 30

The PCL/PDTC block copolymer did not form ring-banded spherulites nor, at 45 °C, did pure PCL, although PCL crystallised from blends at 45 °C often does form ring-banded spherulites (see earlier sections). The pure block copolymer and blends containing 33 wt % of the PCL as homopolymer (i.e. mass of pure PCL/(mass of pure PCL+mass of PCL in copolymer)=0.33) were crystallised at 60 °C (in order to crystallise the PDTC blocks of the copolymer) and dendritic structures were formed. The samples were then cooled to 45 °C to crystallise the PCL blocks and PCL homopolymer. In the pure copolymer the resulting spherulites were coarse and open while in the blend ring-banded spherulites formed. Blends with more than 40 wt % PCL as homopolymer underwent macroscopic phase separation in the melt and, when crystallised, gave ring-banded spherulites. The results were taken to indicate that the PCL homopolymer, of higher molecular weight than the PCL block in the copolymer, phase separated from the copolymer [52]. This phenomenon of homopolymer segregation in mobile systems is well-known in blends of block and graft copolymers with homopolymer of one component where the homopolymer has a molecular weight equal to or greater than the molecular weight of the corresponding block in the copolymer [159].

19.2.2
ε-Caprolactone-Butadiene Diblock Copolymers

Nojima et al. [160] examined the nature of spherulites in blends of PCL with ε-caprolactone-butadiene diblock copolymers. Blends were prepared by solvent-casting mixtures of the components, using a series of block copolymers and PCL samples. Optical microscopy showed that all blends investigated, and cast at

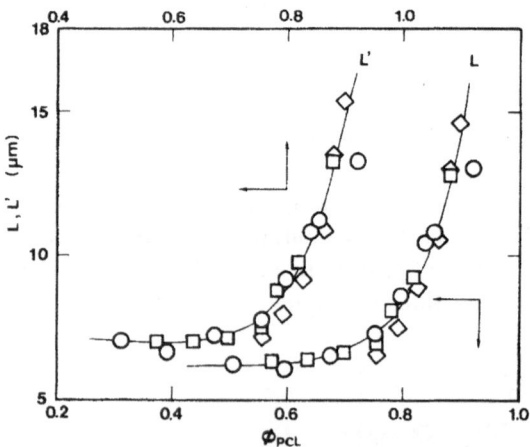

Fig. 70. Variations in long spacing in blends of PCL with ε-caprolactone-butadiene diblock copolymer as functions of homopolymer content for different copolymer compositions: (○) 82 wt % PB, (◊) 61 wt % PB, (△) 35 wt % PB; taken from [160]

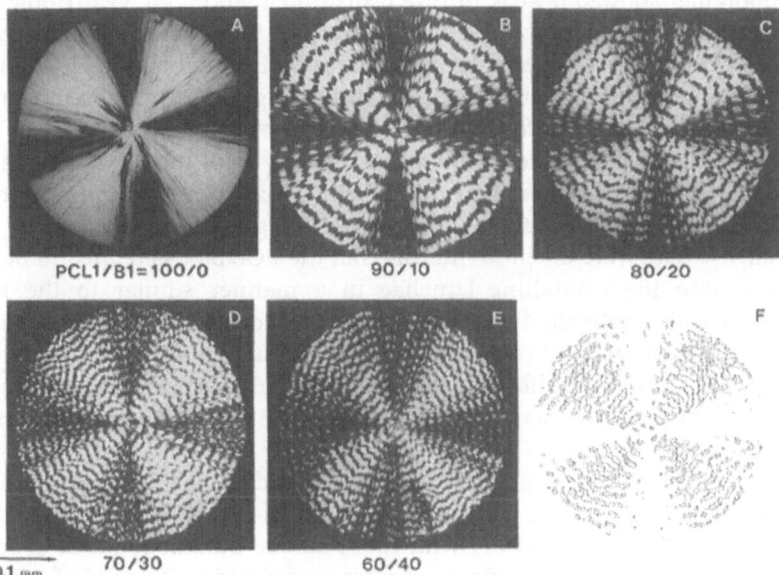

Fig. 71A–F. Ring-banded structures in blends of PCL with ε-caprolactone-butadiene diblock copolymer: **A** pure PCL; **B–E** blends containing 90, 80, 70 and 60% PCL, copolymer contains 82% polybutadiene; **F** simulation; taken from [160]

40 °C, developed ring-banded spherulites which were not apparent in pure PCL cast at 40 °C; crude ring-banding might be seen in PCL crystallised above 45 °C. Ring-banded spherulites are more commonly noted in blends of a crystallisable polymer with another polymer with which it is miscible in the amorphous state; this situation does not apply in this case. The repeat distance of the ring-banding (combined thickness of black and white rings) increased, from about 5 μm to about 15 μm, with increasing volume fraction of PCL homopolymer, increasing more rapidly at high PCL contents (Fig. 70) The periodicity of the ring-banding increased with crystallisation temperature. The ring-banding (Fig. 71) was less regular than seen in some systems. When the data were replotted as a function of total PCL content, then the repeat distances of the ring-banded structures were virtually identical for all blends with the same PCL content. SAXS data from the crystalline blends showed a single scattering peak. The long-spacing calculated from those data decreased linearly with increasing volume fraction of PCL homopolymer. These observations were taken to indicate that the PCL blocks were incorporated into the crystalline regions of the blend and the polybutadiene segments were located in the amorphous interlamellar regions.

Thermal transition data were recorded for blends crystallised at 40 °C; pure block copolymers did not crystallise at 40° C and were crystallised at room temperature. Samples with 50 wt % or more PCL homopolymer showed single crystalline melting endotherms at 55–60 °C, depending on composition. The pure block copolymer showed a weak diffuse transition at about 40° C attributed to poorly crystalline PCL blocks. A sample with 30 wt % block copolymer ($\overline{M}_w=$ 8,600, PCL:PB 65:35) exhibited a broad weak endotherm at about 47 °C and a sharper melting endotherm at about 55 °C, suggesting some form of segregation of the components and the formation of two sets of crystalline lamellae. The overall crystallinity of the blends increased linearly with total PCL content at PCL volume fractions higher than 0.7. Crystallinities were determined from DSC data assuming a heat of fusion for PCL of 136 J g^{-1}.

Overall, the data were taken to indicate that the PCL blocks tended to be incorporated into the crystalline lamellae in a manner similar to the PCL homopolymer. The polybutadiene blocks were thus located in the interlamellar amorphous regions and contributed to the ring-banding. It is not clear if the polybutadiene block was uniformly distributed in the amorphous regions but in the preferred model there was some segregation of PB and PCL within the amorphous regions.

Subsequently, Nojima et al. [161, 162] also studied the morphologies of blends of PCL (PCL1: $\overline{M}_w$=8,300, $\overline{M}_w/\overline{M}_n$=1.4; PCL2: $\overline{M}_w$=8,400, $\overline{M}_w/\overline{M}_n$=1.27; PCL3: $\overline{M}_w$=21,600, $\overline{M}_w/\overline{M}_n$=1.61) with caprolactone-butadiene block copolymers in the melt. The copolymers used were as defined in Table 17; from the compositions given in terms of block volumes, the molecular weights of the PCL blocks were calculated to be about 8200 for B7 and 6000 for B16.

These copolymers were also reported to generate ring-banded spherulites over a wide range of volume fractions of PCL. B16 copolymer alone was found to have a lamellar morphology and B7 was assumed to be lamellar also.

The blends were studied in the melt at 65 °C by SAXS. The maximum of the primary scattering peak became more diffuse with increasing volume fraction of PCL homopolymer ($\phi_{PCL,homo}$). In blends with high $\phi_{PCL,homo}$ an additional diffuse scattering peak was also observable. The repeat distance of the microdomain structure (presumed to be lamellar) of the copolymer in the melt increased with increasing $\phi_{PCL,homo}$ (Fig. 72). The spacing increased more rapidly in blends of B16 with PCL1 than in blends of B7 with PCL2; for the blends of PCL3 with B7 the spacing hardly changed with homopolymer content and it was concluded that these samples exhibited macroscopic phase separation. Using a relation previously developed by Tanaka et al. [163], namely

Table 17. Characteristics of PCL-polybutadiene copolymers used in blends with PCL; data taken from [161]

| Copolymer | $10^{-3}\overline{M}_w$ (total) | $\overline{M}_w/\overline{M}_n$ | vol% B | Polybutadiene microstructure | | | T_m (°C) |
				cis-1,4	trans-1,4	1,2	
B7	12.4	1.10	39	36	52	12	55
B16	11.4	1.18	53	37	51	12	51

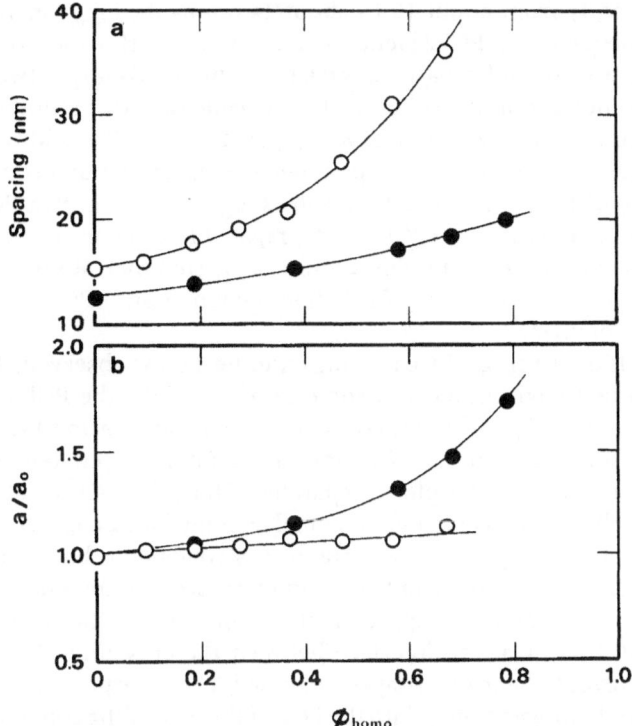

Fig. 72. Variation in lamellar domain spacing in blends of PCL with copolymers B7 and B16, identified in Table 12, with PCL content; taken from [161]

$$\frac{\alpha}{\alpha_0} = \left[\frac{D}{D_0} \left(1 - \phi_{\text{homo}}\right) \right]^{-1/2}$$

which relates phase volumes to spacings in lamellar structures; α is the average distance between chemical junctions of PCL and PB blocks in the blends, α_0 that in the pure copolymer, D_0 is the domain spacing in the pure copolymer and D that in the blend. Nojima et al. [161] found that α/α_0 for B7 blends increased far more rapidly, with increasing $\phi_{PCL,homo}$, than with B16 blends where almost no change in α/α_0 was observed. These results were taken to indicate that PCL swelled the B7 copolymer (i.e. was incorporated into the lamellar structure) but not the B16 copolymer; in the latter case the authors referred to 'localised solubilization'.

The blends were allowed to crystallise at 35 °C and SAXS experiments were performed after crystallisation and long-spacings (L) were determined. The position of the scattering peak intensity shifted to larger s ($s=2sin\theta/\lambda$) with increasing $\phi_{PCL,homo}$ in the blends. A single major scattering peak was observed for B7/PCL1 blends from which long-spacings were determined. For these blends L

decreased linearly from about 19 to about 14 nm as $\phi_{PCL,homo}$ increased from zero to 100 vol%. For B16/PCL2 blends a more complex behaviour was observed. A single peak was found for $\phi_{PCL,homo} < 0.6$ but at higher $\phi_{PCL,homo}$ two scattering peaks were found. From one set of peaks the value of L was found to decrease from about 20 to about 17.5 nm for $\phi_{PCL,homo}$ 0.45 to 0.9. The additional peaks, which were identical for PCL homopolymer, corresponded to a constant long-spacing of about 13 nm. These latter results suggested that in B16/PCL2 blends at high $\phi_{PCL,homo}$ at least some PCL was segregated into separate PCL homopolymer regions within the blend. The authors suggested that, at higher PCL contents, the samples were mosaics of PCL homopolymer and (PCL homopolymer plus block copolymer).

In all blends only a single PCL melting endotherm was observed, despite PCL being in different environments in some samples. While the PCL crystallinity (determined from PCL endotherms) increased with increasing PCL content it remained sensibly constant as a fraction of the total PCL content, even if the sample was macroscopically phase separated. Thus, the PCL in fibrils within spherulites melted similarly to that in lamellae in the block copolymers. From a combination of melting and SAXS data, it was concluded that, as the volume fraction of PCL homopolymer in the blend increased, the thickness of the PCL crystalline lamellae remained approximately constant but the thickness of the amorphous PCL layer at the interface between the crystalline PCL layer and amorphous polybutadiene block layer decreased. The incorporation of the PCL1 homopolymer 'homogeneously' into the PCL of the PCL of the copolymer B7 was recognised as extending and thinning the amorphous polybutadiene layer of the block copolymer [162].

In their previous paper, Nojima et al. [160] suggested that PCL in the block and homopolymer crystallised similarly. They further suggested that the PCL homopolymer and blocks crystallised at the same rates in the B7/PCL1 blends. However, it was concluded that in B16/PCL2 blends PCL crystallisation started simultaneously and independently in pure PCL regions and in the PCL block regions, at least where segregation occurred.

Phase separation of high-molecular-weight PCL from block copolymer in the melt was to be expected [159]. However, the differences in behaviour in these blends were very marked and were attributed to the very small differences which existed in the molecular weights of the homo- and copolymers. Either the system was extremely sensitive to minor differences in block molecular weights etc. in the composition range studied or they were sensitive to some parameter not yet recognised.

20
Ternary Blends

While several possible situations can exist in binary blends, there is even greater potential for complexity in ternary blends. Some options are (i) three components of which all pairs are mutually immiscible, (ii) two components miscible and one immiscible with both, and (iii) two immiscible polymers each of which is miscible with the third in binary blends.

This last type of ternary blend has been studied for PCL with two other polymers which are mutually immiscible but each of which is miscible with PCL. This problem can be considered as one in which the miscible component, in this case PCL, partitions itself between the other components according to the interaction parameters and blend composition. Depending on the relative affinities of PCL for the other components, the PCL will not distribute itself equally between them but will favour one component over the other. This situation is seen in blends of PCL with phenoxy and SAN-15 [136].

Shah et al. [43] investigated blends of bisphenol-A polycarbonate (Dow XP-73009–00, moulding grade) and SAN-25 (Dow, moulding grade) with PCL (PCL-700, Table 1); PC and SAN are mutually immiscible but each is miscible with PCL (Sects. 11 and 15.5). Blends were prepared by melt mixing for 10 min between 220 °C and 270 °C or by solvent casting from solution in dichloromethane, drying samples at room temperature for 24 h and for an additional 24 h at 60 °C under vacuum. Samples were then pressed between 175 °C and 225 °C, according to composition. Most samples were opaque, partly because of PC crystallinity induced by solvent or PCL. Further heterogeneity arose because of phase separation between components which were immiscible in binary mixtures. Blends of SAN with low contents of PCL were transparent.

These workers found single, composition-dependent T_gs for the binary mixtures of PCL with the other components which agreed with the results of previous studies (see earlier sections). Determination of the phase diagram from sample transparency proved difficult as was use of a single T_g as a criterion for miscibility, because of possible overlap of T_gs for individual phases; T_gs which are close are difficult to resolve. Figures 73–75 show T_g data for blends of PCL with SAN-25 and PC in different ratios. From Fig. 73 (SAN-25/PC 1/3 w/w) two T_gs are seen for mixtures low in PCL but, as PCL contents increased, these transitions merged. Thermal transition behaviour varied markedly as the SAN-25/PC ratio was changed. Although the curves for variation of T_g with composition in SAN/PC 3/1 mixtures merged at about 50% PCL, two very distinct T_gs were seen in SAN-25/PC 1/1 blends up to 40 wt % PCL, above which a single T_g was observed. While for samples with SAN-25/PC 3/1 a single T_g was observed at all PCL contents, even as low as 1 wt %; this latter observation strongly suggested that PCL caused true miscibility in these ternary blends with low PC content. On the basis of glass-transition temperature data for SAN-25/PC/PCL blends a ternary phase diagram, based on the observation of a single T_g (Fig. 76) was constructed; this diagram is not a true phase diagram in that it does not recognise separation of crystalline polymer phases.

From crystal melting temperatures, values of B in Eq. (8) were determined, using a value for $v_{2u}/\Delta h_{2u}$ of 134 J cm^{-3} for PCL determined by Ong and Price [93]. From the various data the values of B_{ij} determined were for PCL/SAN –2.6 J cm^{-3}, for PCL/PC –1.6 J cm^{-3} while the value for SAN/PC could not be determined; the uncertainty in this latter value was greater than the mean value of 0.8 J cm^{-3} cal-

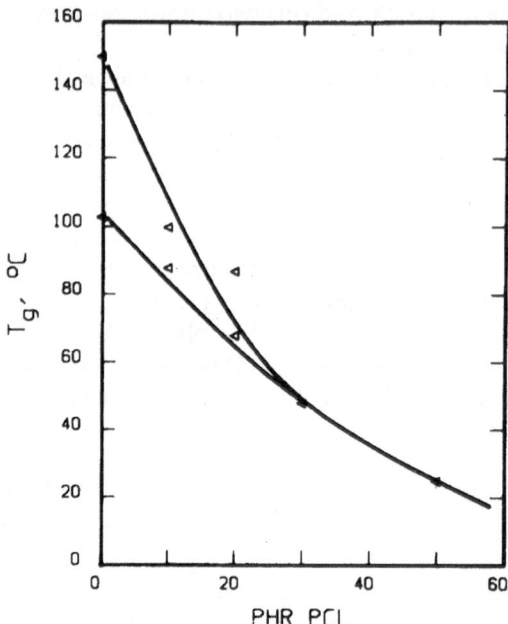

Fig. 73. Variation of glass-transition temperature with PCL content in blends of PCL with SAN-25 and PC (PCL contents defined as parts per hundred of SAN-25 plus PC), SAN-25/PC ratio 1:3; taken from [43]

culated. All the values were small and similar and uncertainties in the values were relatively great. Estimated phase boundaries calculated from the estimated values of the interaction parameters were symmetrical but the experimental phase boundary presented in Fig. 76 was not; this situation suggests that the interaction parameter is not simple and it is necessary to assume a composition-dependent interaction parameter which is inconsistent with simple theory [43].

Blends of PCL ($\overline{M}_n$=15,500) with polycarbonate ($\overline{M}_n$=15,500) and SAN-25 ($\overline{M}_n$=63,000) with SAN/PC 75/25 were also investigated by Larsson and Bertilsson [164]. It was known [165] that PC and SAN are miscible under certain conditions of molecular weight. High-molecular-weight polymers are immiscible but low-molecular-weight polymers, larger than oligomers, are miscible. This contrasts with the observations of Shah et al. [43] who said 1% PCL would induce miscibility. The blends were prepared by melt mixing for 7 min in a Brabender Plasticorder operating at 50 rpm and 250 °C; tensile and impact specimens were injection moulded at 200 rpm and 250 °C; impact specimens were injected into a mould at 50 °C.

Larsson and Bertilsson observed a single T_g for the SAN phase only (at about 100 °C), not for the PC phase and no melting endotherm for PCL for up to 25 wt % PCL [164]; T_g of the SAN phase was broadened. Optical clarity of the blends increased as the PCL content increased and DMTA also showed a broadened loss

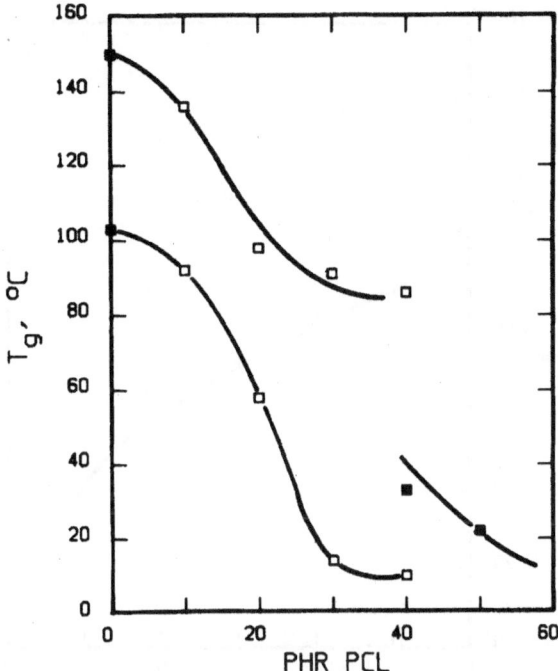

Fig. 74. Variations in glass-transition temperature with PCL content in blends of PCL with SAN-25 and PC (PCL contents defined in legend to Fig. 73), SAN-25/PC ratio 1:1; taken from [43]

peak with development of a shoulder (at 67–70 °C) at the highest PCL content (25 wt %). For the PC phase in pure PC/SAN blend the tanδ peak (1 Hz) was at 158° C (cone and plate rheometry) and addition of a little PCL caused the tanδ peak to shift markedly so that it merged into the SAN peak. Electron microscopy confirmed the presence of two phases in these ternary blends with little change in phase dimensions. In the presence of PCL, fracture surfaces showed less evidence of loss of particles of the dispersed phase. Addition of PCL (3 wt %) to SAN/PC 75/25 blend decreased the tensile modulus (3470 MPa to 3360 MPa) and increased the tensile (48.0 MPa to 62.4 MPa) and impact (19.0 kJ m⁻² to 30.8 kJ m⁻², unnotched) strengths and strain at break (1.5 to 2.2), but all samples underwent brittle fracture. Overall, the results were consistent with immiscibility in the ternary blends to form SAN and PC phases with the PCL partitioned between the two phases and favouring the PC phase; simultaneously the difference in refractive index between the phases was reduced and interfacial adhesion was improved.

Christiansen et al. [166] studied the ternary system PCL/PC/phenoxy in which PC and phenoxy are mutually immiscible; the polymers were the same as used in the previous study by Shah et al. [43] except for the phenoxy (Union Carbide, $\overline{M}_n$=23,000, $\overline{M}_w$=80,000). The procedures used for sample preparation

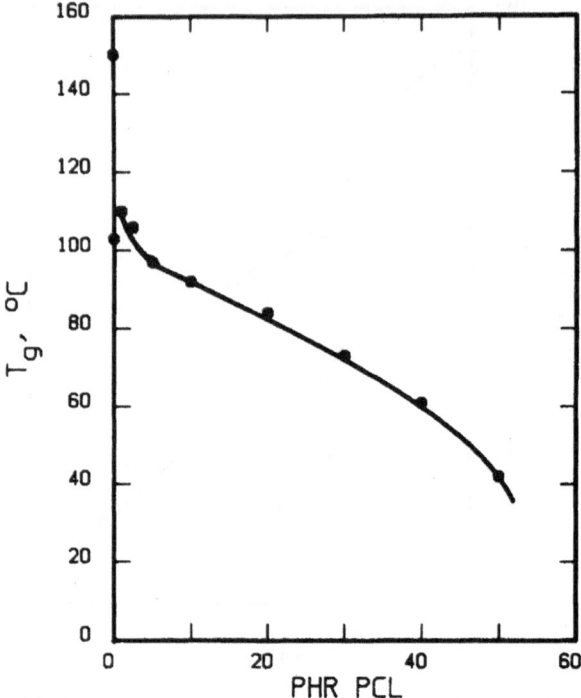

Fig. 75. Variations in glass-transition temperature with PCL content in blends of PCL with SAN-25 and PC (PCL contents defined in legend to Fig. 73), SAN-25/PC ratio 3:1; taken from [43]

were as used for PCL/PC/SAN blends. Binary blends of PC and phenoxy showed two T_gs, equal to those for the pure components, for all compositions. T_gs for the PCL/PC and PCL/phenoxy binary systems were identical with those obtained previously by other workers and described previously. For ternary blends with various ratios of PC/phenoxy (1:3, 1:1 (Fig. 77), 3:1) similar glass-transition behaviour was observed. In all cases two T_gs, from PC and phenoxy-rich phases, were observed as PCL content increased to about 25 wt % PCL, above which only one T_g from PCL/phenoxy mixtures was observed. Blends containing less than 60 wt % PCL also showed the presence of a crystalline PC phase. The greater depression of T_g of phenoxy, compared with that for PC, in the presence of PCL, was taken to indicate that phenoxy had a greater affinity for PCL than did PC. Lack of observance of a T_g from a PC phase as PCL contents increased coincided with the onset of PCL crystallinity and may have been due to overlap of T_g and the T_m for PCL melting. A ternary phase diagram was constructed based on the observation of one or two T_gs and the presence of cloudy melts (Fig. 78). All blends were opaque at temperatures between T_g and PC melting, about 230 °C. Above 230 °C blends containing more than 60 wt % PCL became clear while blends containing 50 wt % PCL remained cloudy.

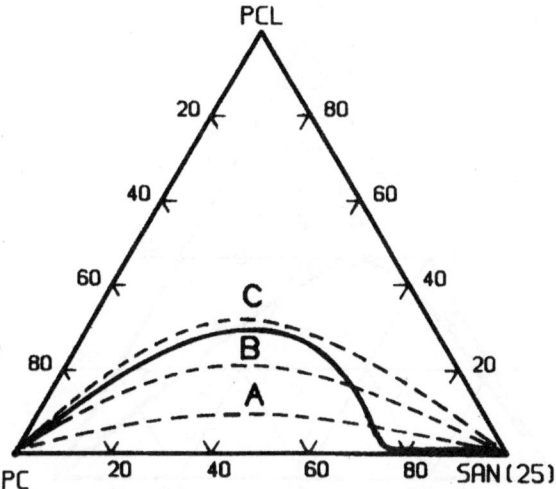

Fig. 76. Experimentally determined boundary of two-phase region (*solid line*) for PCL/PC/SAN-25 blends and boundaries calculated for values of interaction parameters $B_{PCL/SAN}$ –0.61 cal cm^{-3}, B_{PCL} –0.39 cal cm^{-3} and different values of $B_{PC/SAN}$: *curve A* 0.2 cal cm^{-3}, *curve B* 0.5 cal cm^{-3} and *curve C* 0.9 cal cm^{-3}; taken from [43]

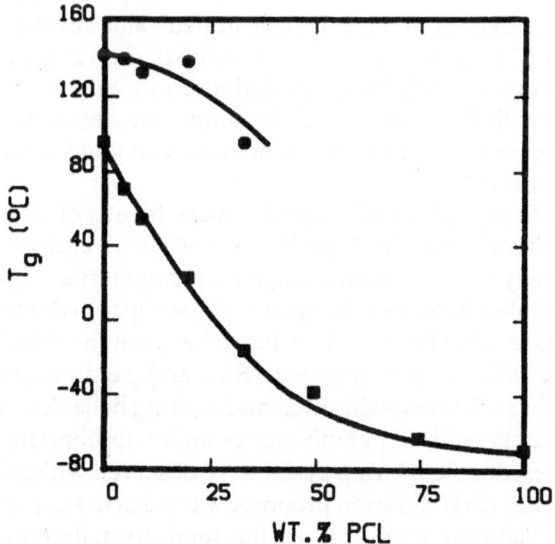

Fig. 77. Variations in glass-transition temperatures with PCL content in ternary blends of PCL with PC and phenoxy (PC/phenoxy ratio (w/w) 1:1), *upper points* are from the PC-rich phase; taken from [166]

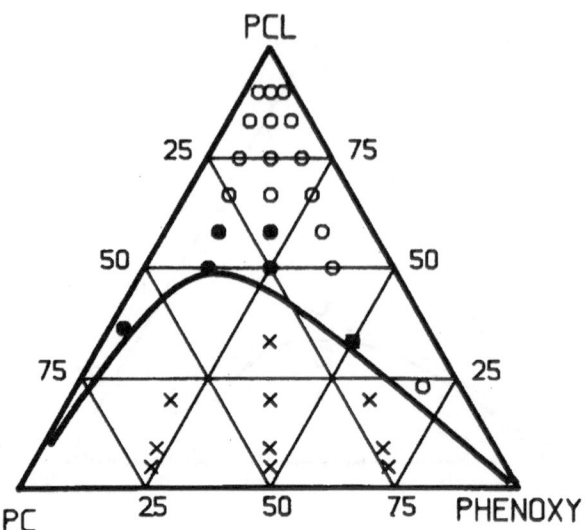

Fig. 78. Ternary phase diagram for PCL/PC/phenoxy blends: (O) single T_g and clear melt, (●) single T_g and cloudy melt, (×) two T_gs and cloudy melts, curve is calculated boundary between single and multiple phases as described in text; taken from [166]

From melting point depressions for PCL melting, Christiansen et al. [166] used the procedures previously applied by Shah et al. [43] to determine interaction parameters. Based on previously determined values of the interaction parameters for PCL/PC, given in Sect. 7, they estimated a value of +19 J cm^{-3} for PC/phenoxy mixtures which indicated endothermic mixing of these components, consistent with PC/phenoxy immiscibility. The several values of interaction parameters gave a calculated phase boundary in good agreement with experimental observations.

Vanneste and Groeninckx investigated ternary blends of PCL with phenoxy and SAN-15. They confirmed the miscibility of PCL with each of the other components individually but demonstrated that only those ternary systems very rich in PCL were miscible. Otherwise the system phase separated into SAN-rich and phenoxy-rich phases with the PCL distributed between them; PCL was found to be more miscible with phenoxy than with SAN and partitioned itself in favour of the phenoxy phase. The crystallisation and melting behaviour of these blends was complex, PCL crystallised in both phases under appropriate circumstances and, in some cases, double melting peaks were observed within a single phase. The mechanisms for crystallisation proposed were discussed [136].

Subsequently, Vanneste et al. studied the semi-crystalline morphologies of blends of PCL ($\overline{M}_n$=38,000, $\overline{M}_w/\overline{M}_n$=1.49) with phenoxy ($\overline{M}_n$=18,000, $\overline{M}_w/\overline{M}_n$=2.59) and SAN-15 ($\overline{M}_n$=71,000, $\overline{M}_w/\overline{M}_n$=2.11) by SAXS [60]. They prepared blends by co-precipitation from solutions in tetrahydrofuran into hexane; samples were dried for 3 days under vacuum at 45 °C. They defined a longspacing L as the distance comprising an amorphous and a crystalline layer,

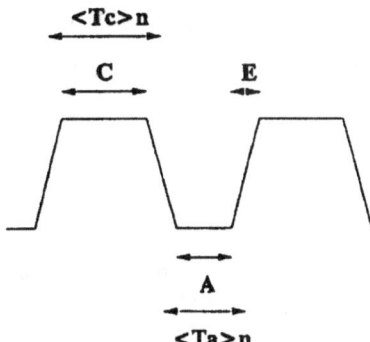

Fig. 79. Pseudo two-phase model showing the variation in density through the alternating crystalline and amorphous layers; thickness C of the crystalline layer is $(<T_c>n-E)$ and thickness A of the amorphous layer is $(<T_a>n-E)$ where E is the thickness of the transition layer; taken from [60]

thickness of a transition layer E between crystalline and amorphous layers as well as average thicknesses of the crystalline and amorphous layers, $<T_c>n$ and $<T_a>n$, respectively (Fig. 79). This investigation was restricted to miscible ternary blends with 90 wt % PCL, as well as pure PCL. The addition of 10 wt % amorphous material to PCL increased values of L slightly, which increased further as the proportion of SAN-15 in the blend (with respect to phenoxy) increased; L increased from 122 Å to 130 Å as phenoxy was replaced by SAN-15. Introduction of 10 wt % amorphous material to PCL decreased the thickness of the crystalline lamellae, relative to PCL; the crystalline thickness increased slightly with increasing proportion of SAN-15. The thickness of the amorphous layer was more sensitive to the nature of the amorphous material; the amorphous layer thickness increased with the SAN-15 content. These various effects, together with SAXS data, were reconciled by proposing that SAN-15 tended to give interlamellar segregation while phenoxy caused interfibrillar segregation, the ternary blends giving intermediate behaviour according to the proportions of SAN-15 and phenoxy in the blend (Fig. 10) [60].

Vanneste et al. also examined the influence of varying the crystallisation temperature (T_c) when it was shown that an increase in T_c increased the long-spacing with a large increase in the thickness of the amorphous layer, attributed to melting of the thinnest lamellae. The crystal layer thickness increased and then decreased as T_c increased from 22 °C to 60 °C. These data correlated with the observation of double melting endotherms which shifted to higher temperature as T_c increased; the lower-temperature endotherm shifted upwards more rapidly. There was also a decrease in overall crystallinity as T_c was raised. The two effects were held responsible for the maximum long-spacing seen at intermediate crystallisation temperatures [60].

Vanneste and Groeninckx also studied blends of PCL with SAN-15 and the copolymer of styrene and maleic anhydride (SMA-14) [149]. These components

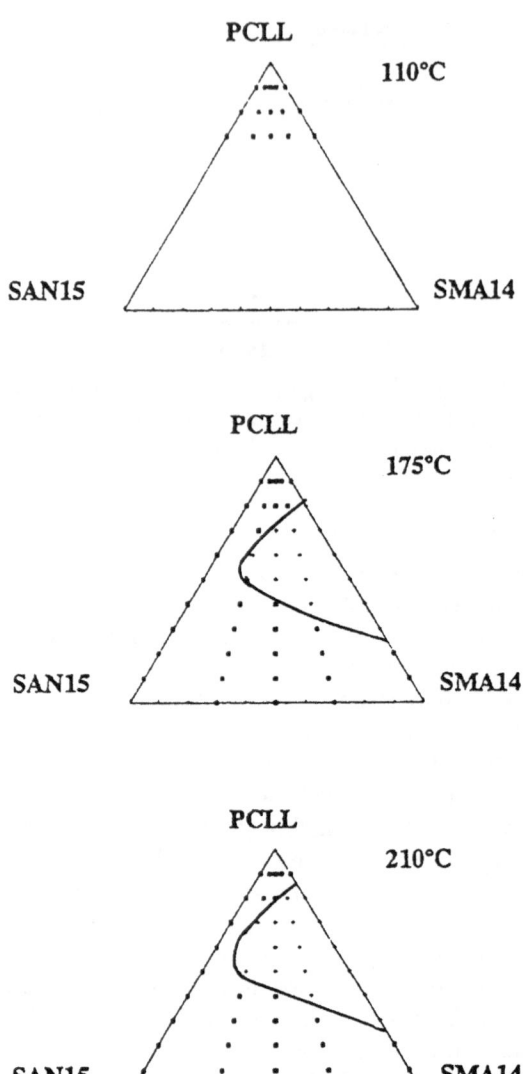

Fig. 80. Miscibility/composition diagrams for ternary PCL (low molecular weight)/SAN-15/SMA-14 blends at different temperatures, molecular weights defined in text; the *solid line* is a suggested phase boundary separating miscible (*larger dots*) and immiscible (*smaller dots*) mixtures; taken from [149]

were chosen because, on the basis of existing evidence, each binary pair was expected to be miscible, at least within certain molecular-weight limits. For PCL and SAN blends there was miscibility of SAN with 6–28 wt % acrylonitrile (AN) (see Sect. 11). While Defieuw et al. reported miscibility between PCL and SMA-14 and SMA-28 (Sect. 17.5) [148], Vanneste and Groeninckx reportedly studied a wider range of compositions and found PCL to be immiscible with SMA-2, to have LCST behaviour with SMA-8 and SMA-14 and to be completely miscible with SMA-17 and SMA-28 [149]. They also commented that others had shown SAN and SMA with similar styrene contents to be miscible. Both binary blends with PCL showed multiple melting peaks in co-precipitated samples [102, 103, 148].

Samples of PCL were of low ($\overline{M}_n$=38,000, $\overline{M}_w/\overline{M}_n$=1.5) and high ($\overline{M}_n$= 53,000, $\overline{M}_w/\overline{M}_n$=1.9) molecular weight. Molecular weight of SAN-15 was $\overline{M}_n$= 65,000 ($\overline{M}_w/\overline{M}_n$=1.9) and SMA-14 $\overline{M}_n$=119,000 ($\overline{M}_w/\overline{M}_n$=4.1). Blends were obtained by co-precipitation of the polymers in THF into hexane. Optical clarity above the melting point was taken to indicate miscibility of the blends. As a result of these studies it was found that blends rich in SAN-15 were all miscible. However, ternary systems with both high and low-molecular-weight PCL showed immiscibility in compositions with less than 50 wt % SAN-15 at 210 °C for high-molecular-weight PCL and less than 40 wt % SAN-15 at 210 °C for low-molecular-weight PCL. Both PCL samples with SMA-14 were immiscible over some composition range at 175 °C and above; at 100 °C most samples were not melted and comparable judgement on miscibility was not possible. The authors proposed phase boundaries for the several systems (Figs. 80 and 81). From these figures it is apparent that overall, for ternary blends, only those with either a little PCL or a little SMA-14 are miscible. Those blends with substantial proportions of PCL and SMA-14 are immiscible. As would be expected, the area of the miscible region is greater for the low-molecular-weight PCL [149].

Because of the partial miscibility of blends of many compositions and consequent complications arising in the positions and breadth of thermal transitions, detailed DSC and dynamic mechanical thermal analysis measurements were restricted to simpler cases. These measurements were restricted to blends with <50 wt % PCL which were the only ones which could be quenched from the melt without the PCL crystallising. For blends with 30 wt % or 40 wt % high-molecular-weight PCL only one T_g was observed.

A consequence of the form of the phase diagrams for these systems is that where immiscible blends formed two phases, only one of those phases contained sufficient PCL for it to crystallise [149]. This result contrasted with the PCL/phenoxy/SAN-15 blends in which two phases formed, one rich in phenoxy and one rich in SAN, each of which contained PCL which could crystallise and resulted in multiple crystal melting endotherms [136].

In this study, Vanneste and Groeninckx investigated crystallisation and melting in blends rich in PCL (90 wt %), from which PCL could crystallise [149]. The presence of SAN-15 and/or SMA-14 reduced the temperature at which PCL could crystallise on cooling from the melt, compared with that for pure PCL. The SAN and SMA components also reduced the rate of crystallisation and crystallisation exotherms were broader than for pure PCL. The presence of the amorphous SAN and SMA also reduced the crystallinity of the PCL which was achieved [149].

Although the ternary blends did not give multiple melting peaks due to the existence of two crystallisable phases, all crystalline blends (with 50 wt % or

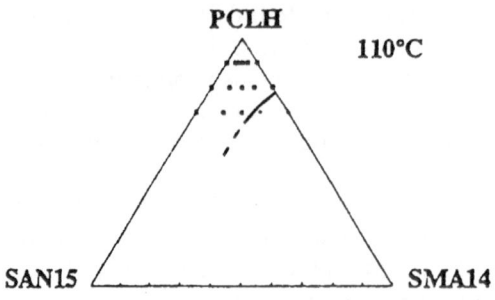

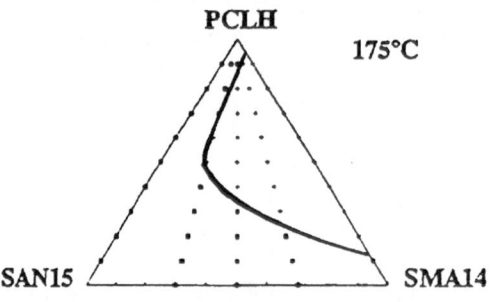

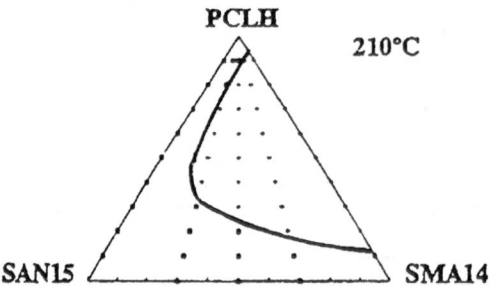

Fig. 81. Miscibility/composition diagrams for ternary PCL (high molecular weight)/SAN-15/SMA-14 blends at different temperatures, molecular weights defined in text; *full squares* denote miscible mixtures and *smaller dots* immiscible mixtures, *solid line* is a suggested phase boundary; taken from [149]

more PCL) examined which had been crystallised at 22 °C for 13 days gave multiple melting endotherms by DSC, a feature attributed to recrystallisation of PCL while heating. With pure PCL there was a single melting endotherm with a tail to low temperature. This tail was more pronounced in a blend with 90 wt % PCL and became a definite peak with less PCL. With 50 wt % PCL the low-temperature peak was the more pronounced. These low-temperature peaks moved from about 45 °C for blends with 80 wt % PCL to about 52 °C for blends with 50 wt % PCL; 60 wt % or more PCL also caused some noticeable decrease in the peak temperature of the upper peak. Crystallisation at higher temperature caused the location of the low-temperature peak to move to higher temperature [149].

With additional data from SAXS studies, Vanneste and Groeninckx determined the thickness of lamellar PCL crystals, their separation and the thickness of the transition layer at the surface of the lamellae. Results suggested that SAN and SMA were located in the interlamellar regions of the high-molecular-weight PCL spherulites. With both high and low-molecular-weight PCL, the thickness of the transition region remained constant with increasing SAN and SMA contents. The amorphous layer thickness increased with SAN and SMA contents. Decreasing the molecular weight of the PCL also reduced the interlamellar thickness and increased the lamellar thickness, and it was concluded that low-molecular-weight PCL gave rise to more-perfect spherulites [149].

Relevant to the study of ternary blends of PCL/phenoxy/PVME are the data on PCL/phenoxy and PCL/PVME binary blends discussed earlier (Sects. 16 and 18.2). Also relevant are studies of phenoxy/PVME blends. Both Robeson et al. [167] and Uriarte et al. [168] demonstrated miscibility over the whole composition range as evidenced by single, composition-dependent T_gs; the latter group also demonstrated that the binary blends are characterised by LCST behaviour with a rather flat cloud-point curve located at about 430–440 °C over most of the composition range. The same workers determined heats of demixing from the magnitudes of relevant endothermic transitions obtained by differential scanning calorimetry.

Qipeng [169] reported on ternary blends of PCL with phenoxy and PVME. PCL/phenoxy and phenoxy/PVME pairs had previously been shown to produce blends characterised by single T_gs; the former blends would have contained some crystalline PCL. The PCL used in this study had $\overline{M}_w$=70,000–100,000, phenoxy had $\overline{M}_w$=43,000 and PVME a viscosity-average molecular weight of 31,000; blends were cast from THF and were dried under vacuum at 50 °C for 4 weeks. In order to minimise the influence of PCL crystallinity blends used for DSC measurements were heated to 97 °C and quenched to –133 °C.

Qipeng demonstrated that PCL/PVME blends were also characterised by single, composition-dependent T_gs (Sect. 18.2). Similarly, all ternary blends of the three polymers exhibited single glass transition temperatures which agreed with values calculated from $1/T_g = \Sigma(w_i/T_{gi})$, an extension of Eq. (23) to multiple components. The author concluded that the polymers were miscible in all proportions but made no reference to the occurrence of PCL crystallisation in the samples. Cloud points were also determined in PCL/PVME and phenoxy/PVME blends; PCL/phenoxy blends remained clear to 200 °C. All ternary systems exhibited cloud points and the minimum of the connecting surface was 108 °C at a PCL/phenoxy/PVME composition of 25/25/50.

Watanabe et al. [82] investigated ternary blends of PCL with polystyrene and poly(vinyl methyl ether). They calculated phase boundaries from interaction parameters determined from binary systems. Calculated phase boundaries were in general agreement with experimentally determined boundaries at constant volume fractions of PCL (10%) and of constant PS (10%) with varying volume fractions of the other components. The data suggested miscibility at the edges of the ternary phase diagrams at various temperatures from 100 °C to 300 °C. At constant PS an LCST was observed at about 160 °C at ϕ_{PVME} about 0.4.

21
Reactive Blends

Blends discussed in preceding sections were based on two or more preformed polymers. One disadvantage of such systems is that commercial melt mixing involves two viscous components. Also, intermolecular interactions, which might aid miscibility, can further enhance the viscosity of the medium. An alternative procedure is to form at least one of the polymeric components in situ. There have been two studies to date which have involved oligomeric materials in generating PCL blends.

In addition to systems discussed in this section, some of the hybrid systems discussed in Sect. 22 are also reactive blend systems.

21.1
Blends with Novolac Resins

PCL ($\overline{M}_n$ 70,000–100,000) has been blended with a novolac (a low-molecular-weight condensation product of phenol and formaldehyde) ($\overline{M}_n$=565); in some samples the novolac was crosslinked with hexamethylenetetramine (HMTA). Initial blends were prepared by solution casting from dichloromethane/ethanol (90/10) mixtures; samples were dried at 50° C under vacuum for 72 h. For curing purposes HMTA was added at 15 wt % relative to the novolac.

Prior to curing, the blends were optically clear above the T_m of PCL and exhibited a single composition-dependent T_g for the novolac, showing the amorphous polymers to be miscible; PCL crystallised from blends containing more than 60 wt % PCL. Miscibility could be partly due to the non-negligible entropy of mixing with the low-molecular-weight novolac. FTIR data also showed that the two components developed intramolecular hydrogen bonding; there was an additional carbonyl stretching vibration for the PCL carbonyl at 1725 cm^{-1} in the blends and this increased in intensity with the novolac content. There was also evidence for hydrogen bonding in the FTIR spectra of the novolac component; from the frequency of the additional band, the strength of the hydrogen bonds developed was deduced to be less than that for self hydrogen bonding in the novolac. These data are consistent with the hydrogen bonding known to occur between the phenolic hydroxyl group of 4-hydroxystyrene and PCL (Sect. 17.3). In the uncured blends the PCL crystallised as ring-banded spherulites [88].

Blends were cured by heating successively at 100 °C, 150 °C and 190 °C for two hours at each stage. As the novolac cured and developed a network structure the components became steadily more immiscible as the novolac molecular weight increased and the entropy of mixing decreased; total separation of the components, however, was restricted by the developing network structure. The melting point of the PCL crystalline regions remained sensibly constant but the extent of crystallinity fell markedly as the novolac content increased; cured blends did not become optically clear above the T_m of PCL. There was a reduction in the strength of intramolecular hydrogen bonding and an apparent decrease in the extent of hydrogen bonding, consistent with reduced miscibility. That is, the cured novolac network became immiscible with, and partially phase separated from, the PCL. In the cured blends the texture of the crystalline PCL changed and in a blend containing 10 wt % novolac the crystalline PCL had a dendritic texture. With increasing novolac content the PCL crystalline regions developed a spherulitic texture within which the novolac network was dispersed. Thus, in the cured samples the components were not phase separated on a large scale, presumably as a result of limited possibilities for segregation as the novolac network structure developed [88].

21.2
Blends with Epoxidised Natural Rubber

PCL ($\overline{M}_n$ approximately 1300–6500) with terminal hydroxyl groups was functionalized to provide telechelic PCL with terminal carboxyl groups (XPCL) **31**. This polymer was blended with epoxidised natural rubber, epoxidised to either 25% or 50% (ENR25 and ENR50, respectively). The two polymers were compounded on a two-roll mill at room temperature to provide a self-curing system. The blends were heated to 180 °C under a pressure of 1000 N cm^{-2}, approximately, to cure them. For comparison, samples of the original PCL **32**, with hydroxyl endgroups were blended with the epoxidised rubber [171].

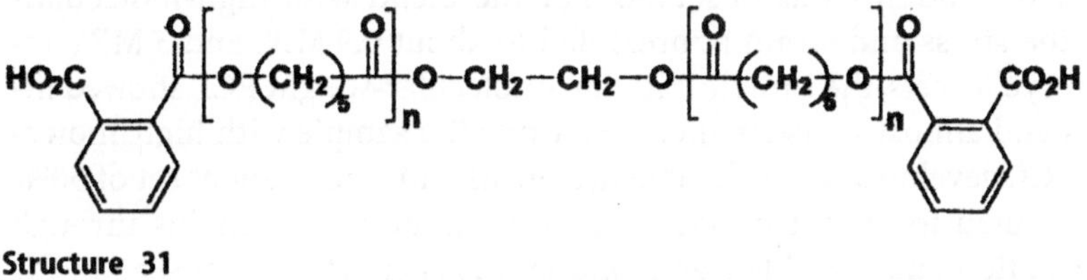

Structure 31

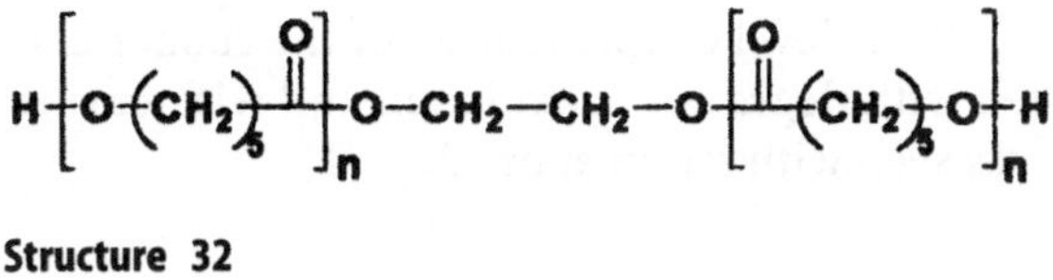

Structure 32

Blends with the original PCL appeared heterogeneous but after curing the reactive blends appeared homogeneous. Rheological measurements during heating showed a large increase in torque associated with curing of the XPCL blends whereas on heating the blends with unfunctionalised PCL the torque did not increase. Thus the XPCL reacted with and crosslinked the epoxidised rubber, and the crosslink density increased with cure time; time-torque curves for blends containing 30 wt % XPCL tended to plateau after heating times of about 1–2 h, depending on the molecular weight of the XPCL, the extent of epoxidation of the rubber and the curing temperature; the more highly epoxidised rubbers developed higher torques at given curing times [171].

After curing, the blends continue to show a T_g for the rubber at about –20 °C, indicative of a phase-separated structure. The T_m peak for the XPCL was shifted to 50–55 °C, endotherm intensities were reduced and for the low-molecular-weight XPCL blend the crystal-melting endotherm was absent; results were consistent with restricted crystallisation of PCL in the cured blends.

Mechanical property data were comparable with those of cured rubbers. With increasing temperature the tensile modulus of ENR50 blends fell in two stages, from about 2 GPa at low temperatures to about 5 MPa just below 0 °C (the T_g of the rubber component) and to 1 MPa above the T_m of PCL. The composite with the low-molecular-weight PCL retained a slightly higher modulus at high temperature and did not show a drop at T_m of PCL, indicating that the low-molecular-weight PCL did not segregate in the cured polymer. The stress-strain curves at room temperature showed that the modulus increased with the molecular weight of the PCL, as did the stress and strain at break. The sample with high-molecular-weight PCL showed typical strain hardening and broke at a stress of 11 MPa and a strain of 6.5 and there was possibly strain-induced crystallisation of the ENR at large strains; other samples failed at strains less than 4 and stresses below 3 MPa. The crystalline high-molecular-weight PCL acted as a filler as well as a crosslinking agent in the rubber. In unstrained samples, electron microscopy of the blend with high-molecular-weight PCL showed the presence of fine PCL crystalline fibrils in the rubber matrix. At 80 °C, i.e. above T_m of PCL, the reverse order of modulus was observed. For the blend with high-molecular-weight PCL the stress and strain at break fell to about 0.9 MPa and 3 MPa, respectively. In cyclic tests, the sample with low-molecular-weight PCL showed little hysteresis and almost no permanent set, while the sample with high-molecular-weight PCL developed considerable hysteresis and a permanent set of 50%, an effect attributed to modifications to the crystalline PCL domains through plastic deformation and possible fragmentation. This effect was not seen in the sample with low-molecular-weight PCL as no crystalline PCL domains were present. The form of the stress strain curves for cured samples with high-molecular-weight PCL was independent of the level of epoxidation of the rubber and the curing time; moduli increased with degree of epoxidation and with curing time, and the reverse tendency was seen with strain at break [171].

21.3
Blends with Unsaturated Polyester Resin

PCL ($\overline{M}_n$=70,000–100,000) has been blended with uncured, unsaturated polyester resin which was subsequently cured [170]. The unsaturated oligomeric resin ($\overline{M}_n$=1000) was a copolyester formed from isophthalic acid, fumaric acid and propylene glycol (in the molar ratio 1:1:2.2) and contained 33 wt % styrene monomer. The resins were cured by adding benzoyl peroxide and heating, in stages, from 60 °C to 120 °C.

Blends of PCL with the uncured resin exhibited a single T_g which in non-crystalline samples, or samples with <4% PCL crystallinity (<60 wt % PCL), fitted Eq. 23. A sample with higher crystallinity had a single T_g commensurate with Eq. 23 when the crystalline content was allowed for in calculating the composition of the amorphous phase; PCL crystallisation temperatures decreased from 40 °C to about –10 °C as the PCL content increased from 30 wt % to 60 wt % PCL. Crystalline samples exhibited double melting peaks attributed to initial imperfect crystallisation and recrystallisation during heating. Above T_m all samples were optically clear. FT-IR studies indicated specific hydrogen-bonding interactions between hydroxyl groups in the resin and carbonyl groups of the PCL which are stronger than self-association in the resin itself. The normal form of PCL spherulites were found in the uncured blends containing higher contents of PCL.

When cured, blends containing >10 wt % PCL were opaque and remained so at elevated temperatures; the blends had phase-separated on curing. Blends with 5 wt % or less PCL remained miscible, or at least miscibility was trapped on curing as crosslinking and network formation developed. On heating cured samples, cooled from the curing temperature, containing 10–30 wt % PCL in a DSC experiment, crystallisation exotherms were observed. At higher PCL contents only crystal melting endotherms were observed; values of T_m for PCL are only slightly reduced from that of pure PCL. FT-IR data demonstrated that hydrogen-bonding interactions exist between PCL and the cured resin in a mixed amorphous phase. Cured blends with higher PCL contents (≤50 wt %) show the presence of ring-banded spherulites.

22
Blends with Inorganic-Organic Hybrid Materials

In addition to polymer-polymer blends, there is interest in inorganic-modified polymers in the form of inorganic-organic hybrids. Fillers are often added to polymers to reduce costs and their influences on properties are often modest. Alternatively an aim can be to prepare polymer-ceramic composites with particular characteristics.

22.1
Montmorillonite-Poly(ε-caprolactone) Hybrids

Messersmith and Giannelis prepared nanocomposites using mica-type silicates [172]. Mica-type silicates are 2:1 layered silicates in which two silica tetrahedral sheets are fused to an edge-shared octahedral sheet of aluminium or magnesium hydroxide. The sheets have a high aspect ratio. Mica-type silicates contain inorganic ions which can be exchanged with organic cations to render them organophilic, allowing them to be intercalated with organic materials. Composites were prepared by modifying sodium montmorillonite with a protonated amino acid, dispersing it in caprolactone and polymerising the caprolactone in situ. Composites containing up to 20 wt % (8 vol.%) mica were prepared. Subsequent breakdown of the composites showed the PCL to have molecular weights ($\overline{M}_n$) in the range 8000 to 17,000. The PCL in the composites was crystalline with its normal crystal structure, but with broadened diffraction peaks as a result of reduced crystallite size. The composites were stable when dispersed in solvents for PCL, such as toluene, as a result of the method of synthesis which resulted in strong interactions between the polymer and the silicate surface. The main interest was in materials with reduced permeability to water; 4.8 wt % silicate (by volume) reduced water uptake by an order of magnitude compared with pure PCL. The effect was attributed to the large aspect ratio of the silicate and the tortuosity of paths through the composite [172].

Jimenez et al. also investigated blends of PCL reinforced by an organically-modified montmorillonite clay, with a view to using inert inorganic structures to enhance the properties of PCL for which T_g and T_m are low. With the montmorillonite having a layered structure, it was hoped to intercalate the polymer from solution with the clay [173].

To produce organically-modified montmorillonite, the montmorillonite clay was mixed in water with distearyldimethylammonium chloride. The resulting solid was extracted and washed with sodium chloride solution and ethanol, to remove excess ammonium salt, and freeze dried. To combine with the PCL, the organically-modified montmorillonite was dispersed in chloroform; the unmodified montmorillonite would not disperse in this way. Samples with up to 30 wt % clay were produced.

T_m and ΔH_m for the PCL decreased with increasing content of the modified montmorillonite in the blends. ΔH_m for the composite varied closely to that expected if the fractional crystallinity of the PCL was constant as its weight fraction decreased. That is, fractional PCL crystallinity appeared to be independent of clay content. Dimensions of PCL spherulites in the presence of the clay were smaller than in pure PCL. Crystallisation studies showed that, relative to the rate of crystallisation in pure PCL, 5 wt % clay enhanced the rate of crystallisation but above 10 wt % clay the rate of crystallisation was decreased. It was proposed that clay enhanced the rate of nucleation but high clay contents hindered diffusion of polymer segments and, hence, of crystallisation.

Small-angle and wide-angle X-ray diffraction studies provided information on the structure of the montmorillonite-PCL blends. The montmorillonite has an inherent, anisotropic layered structure. Data suggested that in the blends the clay layers formed units of stacked layers (tactoids), parallel to the film surface. The PCL crystallites were oriented parallel to the clay but randomly orientated in the film layer. This implies that the PCL molecules in the crystallites had their chain axes perpendicular to the clay surface. It was suggested that at the relatively high temperature used in film formation (50 °C) the clay might influence the chain orientation and that crystallites formed on the clay surface. In the plane of the film, the clay tactoids formed a superlattice with the PCL lamellae between the parallel tactoids. Results indicated that stacks of up to three silicate lamellae form the tactoids which were interleaved with PCL crystallites. It was suggested that the assembly of orientated tactoids could act as a barrier to water diffusion through the films.

Incorporation of modified montmorillonite into the PCL enhanced the draw ratio attainable from 3.3 for pure PCL to 5.6 with 20 wt % clay; with 30 wt % clay the samples exhibited brittle fracture. That is, clay layers (up to 20 wt %) enhance ductility of the blends. Experiments were performed on drawn films. Results suggested that orientation of tactoids was enhanced by drawing. This orientation was apparently associated with formation of voids and of fibrillation between the tactoids which permitted high draw ratios to be attained.

Measurements of dynamic mechanical properties showed that the storage modulus, at any given temperature, was higher in the presence of the montmorillonite; samples showed a gradual decrease with increasing temperature and a distinct decrease at about –60 °C. This step corresponded to a maximum in loss modulus at the same temperature and a maximum in tanδ at –54 °C, which corresponds to T_g of the PCL; there was a slight increase in T_g with increasing volume fraction of montmorillonite. A low-temperature transition at –100 °C corresponded to development of motions of the methylene units.

The small increase in T_g with volume fraction of montmorillonite indicated that interactions between clay and PCL are relatively weak and that PCL and montmorillonite did not fully intercalate. These interactions appeared to collapse at about 35 °C, corresponding to a peak in loss modulus associated with relaxation in the PCL chains [173].

22.2
Silica-Poly(ε-caprolactone) Hybrids

Another type of inorganic-organic hybrid is the ceramers formed from a combination of organic polymer units and components of inorganic glasses, usually combined in a sol-gel process. These materials offer the potential of combining the properties of inorganic glasses, such as optical clarity, thermal stability and high modulus with polymer flexibility. They are prepared by reacting a preformed polymer, or oligomer, having reactive functional groups (usually endgroups) with an inorganic reagent under conditions such that the inorganic re-

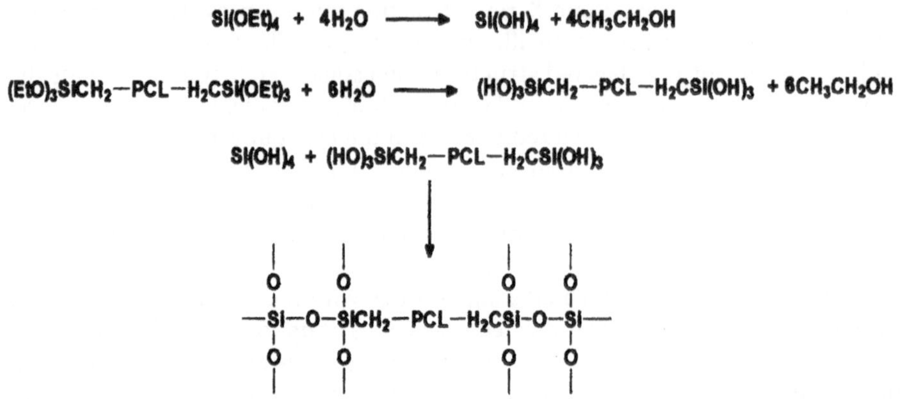

$$Si(OEt)_4 + 4H_2O \longrightarrow Si(OH)_4 + 4CH_3CH_2OH$$

$$(EtO)_3SiCH_2-PCL-H_2CSi(OEt)_3 + 6H_2O \longrightarrow (HO)_3SiCH_2-PCL-H_2CSi(OH)_3 + 6CH_3CH_2OH$$

$$Si(OH)_4 + (HO)_3SiCH_2-PCL-H_2CSi(OH)_3$$

Scheme 1

agent forms a glassy network (gel) structure; the organic component is chemically incorporated into that network structure.

A series of investigations have now been carried out into the synthesis of ceramers incorporating PCL as the organic component and silica as the inorganic component [174, 175]. Tetraethoxysilane (TEOS) was used as the precursor for the inorganic component. This reagent can be hydrolysed and condensed to form a silica network. PCL can be obtained with hydroxyl endgroups (i.e. α,ω-hydroxyl PCL) which can be converted to triethoxysilane groups. Either of these components can be used in the sol-gel process; the use of trifunctional reactive species has also been described. The TEOS and functionalized PCL (typically 1.5 g TEOS and 0.5 g PCL) were reacted together in THF solution with water and mineral (hydrochloric) acid catalyst in the presence of added ethanol; ethanol is a reaction product but its presence was reported to enhance transparency in the final product. The water, with acid, hydrolysed the ethoxy groups to hydroxyl and brought about the condensation reaction which combined the components and caused network formation, i.e. gelation; α,ω-hydroxyl PCL reacted more slowly in polycondensation than chains originally end-capped with triethoxysilane groups. In the case of triethoxysilane end-capped PCL, the overall reaction sequence and gel formation is schematically represented by the reaction scheme (Scheme 1) where the incomplete bonds on the oxygen atoms indicate bonds to other Si or PCL moieties to form three-dimensional networks. Once network formation was complete, careful removal of volatiles converted the gel into optically clear glass without cracks [174, 175].

It has been reported that the PCL in the final ceramer was intimately incorporated into the network structure and remained totally amorphous, consistent with optical clarity. Thus, no phase separation occurred during ceramer formation. Samples with >60 wt % PCL in the final ceramer were reported to be translucent or opaque. Subsequent Soxhlet extraction with THF demonstrated that all PCL was not incorporated into the final network structure. Network constraints and vitrification reduced molecular mobility necessary for continued reaction [175].

High-temperature curing (100 °C for 1 day) enhanced the incorporation of PCL, typically from 21% to 45%. Larger proportions of PCL used were incorporated at low (15%) PCL contents; 41% was incorporated when cured at 25 °C and up to 100% when cured at 100 °C. At higher PCL contents (54%) these proportions were 22% and 80%, respectively. PCL of higher molecular weight was incorporated more effectively. Even PCL without reactive endgroups was partially permanently incorporated into the networks, an effect attributed to hydrogen-bonding between the network and the carbonyl of PCL [175].

Although the samples were optically clear, evidence for a glass transition at low temperatures, attributed to PCL and raised to –55 °C, was observed in the ceramer, implying lack of total homogeneity in the ceramer structure. High-temperature curing for a day enhanced the thermal stability of the PCL within the ceramer but prolonged curing reversed that trend [175]. Subsequent dynamic mechanical analysis showed the presence of relaxation peaks below 0 °C for several monomers but no further loss peaks below 150 °C; details varied with curing conditions, composition and PCL end-group functionality. The loss peaks were associated with a loss of storage modulus from 5 GPa to 1 GPa. SAXS data also indicated microphase segregation of the components on a length scale of about 4 nm [176].

Transmission electron microscopy of samples with <30 wt % PCL showed no evidence of segregation of components. However, samples with higher PCL contents did show evidence of fine-scale segregation. The results of image analysis were consistent with a ceramer with 46 wt % PCL having co-continuous microphases [176].

It was also reported that the contact angle with water and the ceramers increased with PCL content, from 29° at 15% PCL to 78° for pure PCL [177]. Fibroblasts were used to assess the biosensitivity of the ceramers and it was found that the density of fibroblasts adhering to the ceramer surface decreased with increasing PCL content, i.e. as the hydrophilicity decreased. The surface morphology of the ceramers was investigated by atomic force microscopy [177]. The images were interpreted to indicate that ceramers with 50% PCL had dispersed particles (10–15 nm) at the surface; for ceramers with higher PCL contents no surface structures were resolved. In vitro biodegradation at 37 °C of the ceramers containing 50% PCL was studied by immersing them in a sodium phosphate (0.1 mol l^{-1}) buffer (pH 8) containing porcine esterase (23 units cm^{-3}). After 4 days immersion the initial surface morphology had disappeared and the surface composition changed; the PCL content at the surface decreased by about 35%. These changes continued with further immersion and the surfaces became granular. Thus, selective degradation of the PCL was considered to occur with PCL chains, and some silanol units, being lost. Potential applications of the ceramers in biomaterials area were considered [177].

Acknowledgements. The author wishes to thank Solvay Interox, and Dr R. Wasson in particular, who instigated the preparation of this review.

References

1. Mullins DH (1966) Plasticization of vinyl resins with lactone polyesters. Union Carbide Corp assignee. Canadian Pat 742,294
2. Koleske JV, Lundberg RD (1969) J Polym Sci A-2 7:795
3. Kim W (1989) Korean Pat 9,304,290
4. Foster T, Gratowski M, Konstandreas A (1994) Canadian Pat 2,071,078
5. Daicel Chem Ind (assignee) (1992) Transesterification of PCL and polyester or polycarbonate to produce antielectrostatic compounds. Japanese Pat 6,093,252-A
6. Rossmy G, Spiegler R, Venzmer J (1992) German Pat DE4,206,191
7. Cavallaro P, Immirizi B, Malinconico M, Martuscelli E, Volpe MG (1994) Macromol Chem, Rapid Commun 15:103
8. Cavallaro P, Immirzi B, Malinconico M, Martuscelli E, Volpe MG(1993) Angew Makromol Chem 210:129
9. Russell TP, Stein RS (1983) J Polym Sci: Polym Phys Ed 21:999
10. Seefried CGJ, Koleske JV (1974) J Macromol Sci-Phys B10:579
11. Bittiger H, Marchessault RH, Niegisch WD (1970) Acta Crystallographica B26:1923
12. Chatani Y, Okita Y, Tadokoro H, Yamashita Y (1970) Polym J 1:555
13. Crescenzi V, Manzini G, Calzolari G, Borri C (1972) Eur Polym J 8:449
14. Ong CU (1973) Thesis, Univ of Massachusetts
15. Coleman MM, Zarian J (1979) J Polym Sci: Polym Phys Ed 17:837
16. Khambatta FB, Warner F, Russell T, Stein RS (1976) J Polym Sci: Polym Phys Ed 14:1391
17. Shur YJ, Ranby B (1977) J Macromol Sci – Phys B14:565
18. Olabisi O (1975) Macromolecules 8:316
19. Chiu S-C, Smith TG (1984) J Appl Polym Sci 29:1781
20. Chiu S-C, Smith TG (1984) J Appl Polym Sci 29:1797
21. Cruz CA, Paul DR, Barlow JW (1979) J Appl Polym Sci 23:589
22. Tanaka H, Nishi T (1985) Phys Rev Lett 55:1102
23. Barron CA, Kumar SK, Runt JP (1995) Polym Prepr Amer Chem Soc Polym Div 33(2):610
24. Freeman M, Manning PP (1964) J Polym Sci, Part A 2:2017
25. Huggins ML (1941) J Chem Phys 9:440
26. Huggins ML (1942) J Phys Chem 46:151
27. Flory PJ (1941) J Chem Phys 9:660
28. Flory PJ (1942) J Chem Phys 10:51
29. Flory PJ (1953) Principles of polymer chemistry. Cornell Univ Press, Ithaca, NY
30. Scott RL (1949) J Chem Phys 17:279
31. Tompa H (1949) Trans Far Soc 45:1142
32. Paul DR, Newman S (eds) (1978) Polymer blends. Academic Press, San Diego, London
33. Olabisi O, Robeson LM, Shaw MT (1979) Polymer-polymer miscibility. Academic Press, San Diego, London
34. Walsh DJ, Rostami S (1985) Interactions in polymer blends. In: Advances in polymer science, vol 70. Springer, Berlin Heidelberg New York, p 119
35. Coleman MM, Graf JF, Painter PC (1991) Specific interactions and the miscibility of polymer blends. Technomic, Lancaster, USA
36. Rostami S (1992) Polymer-polymer blends. In: Miles IS, Rostami S (eds) Multicomponent polymer systems. Longman, Harlow, UK, p 63
37. Sanchez IC (1978) Statistical thermodynamics of polymer blends. In: Paul DR, Newman S (eds) Polymer blends. Academic Press, San Diego, p 115

38. Olabisi O, Paul DR, Shaw MT (1997) Methods for determining polymer-polymer miscibility. In: Polymer-polymer miscibility. Academic Press, San Diego, p 117
39. McMaster LP (1973) Macromolecules 6:760
40. Nishi T, Wang TT (1975) Macromolecules 8:909
41. Hoffman JD, Weeks JI (1962) J Res Nat Bur Standards A, Chem Phys 66A:13
42. Lezcano EG, Salom Coll C, Prolongo MG(1996) Polymer 37:3603
43. Shah VS, Keitz JD, Paul DR, Barlow JW (1986) J Appl Polym Sci 32:3863
44. Hildebrand JH, Scott RL (1950) The solubility of nonelectrolytes, 3rd edn. Van Nostrand-Reinhold, Princeton
45. Small PA (1953) J Appl Chem 3:71
46. Hoy KL (1970) J Paint Technol 42:76
47. Aubin M, Prud'homme RE (1980) Macromolecules 13:365
48. Braun JM, Guillet JE (1973) In: Purnell JH (ed) Progress in gas chromatography. Wiley (Interscience), New York, p 107
49. Guillet JE (1973) In: Purnell JH (ed) Progress in gas chromatography. Wiley (Interscience), New York, p 187
50. Su SU, Patterson D, Schreiber HP (1976) J Appl Polym Sci 20:1025
51. Nojima S, Watanabe K, Zheng Z, Ashida T (1988) Polym J 20:823
52. Kummerlöwe C, Kammer HW (1995) Polym Networks Blends 5:131
53. Wang Z, An L, Jiang B, Wangh X (1998) Macromol Rapid Commun 19:131
54. Jordhamo GM, Manson JA, Sperling LH (1986) Polym Eng Sci 26:517
55. Kelley FN, Bueche F (1961) J Polym Sci 50:549
56. Gordon M, Taylor JS (1952) J Appl Chem 2:493
57. Fox TG (1956) Bull Amer Phys Soc 2:123
58. Hubbell DS, Cooper SL (1977) J Appl Polym Sci 21:3035
59. Defieuw G, Groeninckx G, Reynaers H (1989) Polymer 30:2164
60. Vanneste M, Groeninckx G, Reynaers H (1997) Polymer 38:4407
61. Hammer CF (1978) Polymeric plasticizers. In: Paul DR, Newman S (eds) Polymer blends. Academic Press, San Diego, p 219
62. Gardiewski N, Schmidt F (1992) Internally plasticized PVC by polymerization of vinyl chloride by polycaprolactone. European Pat 508,057
63. Daicel Chem Ind KK (assignee) (1987) Plastic clay for statue modelling. Japanese Pat J-01,143,457
64. Daicel Chem Ind KK (assignee) (1988) Plastic clay for statue modelling. Japanese Pat J-01,213,353
65. Daicel Chem Ind KK (assignee) (1988) Plastic clay. Japanese Pat J-01197548
66. Daicel Chem Ind KK (assignee) (1988) Plastic clay. Japanese Pat J-01,196,090
67. Bussink J, Sederel WL (1984) European Pat 118,706
68. Garton A (1991) Infrared spectroscopy of polymer blends, composites and surfaces. Hanser, Munich
69. Pouchly J, Biros J (1969) J Polym Sci: PolymLetts 7:463
70. Varnell, DF (1982) Thesis, Pennsylvania State Univ
71. Prud'homme RE (1982) Polym Eng Sci 22:90
72. Riedl B, Prud'homme RE (1986) J Polym Sci B: Polym Phys 24:2565
73. Woo EM, Barlow JW, Paul DR (1985) Polymer 26:763
74. Zhang H, Prud'homme RE(1987) J Polym Sci B: Polym Phys 25:723
75. Higashida N, Kressler J, Inoue T (1995) Polymer 36:2761
76. Higashida N, Kressler J, Yukioka S, Inoue T (1994) Macromolecules 27:2448
77. Svoboda P, Kressler J, Ougizawa T, Inoue T, Ozutsumi K (1997) Macromolecules 30:1973
78. Kressler J, Svoboda P, Inoue T (1992) Polym Prep Amer Chem Soc Polym Div 33(2):612
79. Kressler J, Svoboda P, Inoue T (1993) Polymer 34:3225
80. Li W, Yan R, Jing X, Jiang B (1992) J Macromol Sci – Phys B31:227
81. Li Y, Stein M, Jungnickel B-J (1991) Colloid Polym Sci 269:772

82. Watanabe T, Fujiwara Y, Sumi Y, Nishi T (1982) Rep Progr Polym Phys Japan 25:285
83. Harris JE, Goh SH, Paul DR, Barlow JW (1982) J Appl Polym Sci 27:839
84. De Juana R, Etxeberria A, Cortazar M, Iruin JJ (1994) Macromolecules 27:1395
85. Coleman MM, Moskala EJ (1983) Polymer 24:251
86. Fernandes AC, Barlow JW, Paul DR (1986) Polymer 27:1799
87. Kim CK, Paul DR (1994) Polym Eng Sci 34:24
88. Zhong Z, Guo Q (1997) Polymer 38:279
89. Moskala EJ, Varnell DF, Coleman MM (1985) Polymer 26:228
90. Lezcano EG, Prolongo MG, Salom Coll C (1995) Polymer 36:565
91. Sanchis A, Prolongo MG, Salom Coll C, Masegosa RM (1998) J Polym Sci B: Polym Phys 36:95
92. Barnum RS, Goh SH, Paul DR, Barlow JW (1815) J Appl Polym Sci 26:3917
93. Ong CJ, Price FP (1978) J Polym Sci Polym Symp 63:45
94. Ong CJ, Price FP (1978) J Polym Sci Polym Symp 63:59
95. Nojima S, Tsutsui H, Urushihara M, Kosaka W, Kato N, Ashida T (1986) Polym J 18:451
96. Ajji A, Renaud MC (1991) J Appl Polym Sci 42:335
97. Pingping Z, Haiyang Y, Shiqiang W (1998) Eur Polym J 34:91
98. Ziska JJ, Barlow JW, Paul DR (1981) Polymer 22:918
99. Defieuw G, Groeninckx G, Reynaers H (1989) Polymer Commun 30:267
100. Defieuw G, Groeninckx G, Reynaers H (1989) Polymer 30:595
101. Alberada van Ekenstein GOR, Deuring H, ten Brinke G, Ellis TS (1997) Polymer 38:3025
102. Rim PB, Runt JP (1983) Macromolecules 16:762
103. Runt J, Rim PB (1982) Macromolecules 15:1018
104. Rim PB, Runt JP (1984) Macromolecules 17:1520
105. Rim PB, Runt J (1985) J Appl Polym Sci 30:1545
106. Kressler J, Kammer HW, Silvestre C, Di Pace E, Cimmino S, Martuscelli E (1991) Polym NetworksBlends 1:225
107. Wang Z, Wang X, Yu D, Jiang B (1997) Polymer 38:5897
108. Wang Z, An L, Jiang B, Wang X, Zhao H (1998) Polymer J 30:206
109. Schulze K, Kressler J, Kammer HW (1993) Polymer 34:3704
110. Wang Z, Jiang B (1997) Macromolecules 30:6223
111. Svoboda P, Kressler J, Chiba T, Inoue T, Kammer HW (1994) Macromolecules 27:1154
112. Rocha R, Gross R, McCarthy S (1995) Polym Prep Amer Chem Soc Polym Div 33(2):454
113. Kumagai Y, Doi Y (1992) Polym Degradation Stability 36:241
114. Gassner F, Owen AJ (1994) Polymer 35:2233
115. Yasin M, Tighe BJ (1992) Biomaterials 13:9
116. Stevels WM, Bernard A, Van de Witte P, Dijkstra PJ, Feijen J (1996) J App Polym Sci 62:1295
117. In't Veld PJA, Velner EM, Van de Witte P, Hamhuis J, Dijkstra PJ, Feijen J (1997) J Polym Sci A: Polym Chem 35:219
118. Yang J-M, Chen H-L, You J-W, Hwang JC (1997) Polym J 29:657
119. Ketelaars AAJ, Papantoniou Y, Nakayama K (1997) J Appl Polym Sci 66:921
120. Koleske JV, Whitworth CJ Jr, Lundberg RD (1975) Polyesters blended with cyclic ester polymers. US Pat 3,892,821
121. Vazquez-Torres H, Cruz-Ramos CA (1994) J Appl Polym Sci 54:1141
122. Han K, Kang H-J (1996) Polymer (Korea) 20:224
123. Dezhu M, Xiaolie L, Ruiyun Z, Nishi T (1996) Polymer 37:1575
124. Dezhu M, Xiang X, Xiaolie L, Nishi T (1997) Polymer 38:1131
125. Mercier JP, Groeninckx G, Lesne M (1967) J Polym Sci C 16:2059
126. Jonza JM, Porter RS(1986) Macromolecules 19:1946
127. Hatzius K, Li Y, Werner M, Jungnickel B-J (1996) Angew Makromol Chem 243:177
128. Brandrup J, Immergut EH (eds) (1919) Polymer handbook, 3rd edn. Wiley Interscience, NY

129. Cruz CA, Barlow JW, Paul DR (1979) Macromolecules 12:726
130. Van Krevelen DW (1990) Properties of polymers. Elsevier, Amsterdam
131. Shuster M, Narkis M, Siegmann A (1994) J Appl Polym Sci 52:1383
132. Smith WA, Barlow JW, Paul DR (1981) J Appl Polym Sci 26:4233
133. Robard A, Patterson D, Delmas G (1977) Macromolecules 10:706
134. Brode GL, Koleske JV (1972) J Macromol Sci – Chem A6:1109
135. Espi E, Iruin JJ (1991) Macromolecules 24:6458
136. Vanneste M, Groeninckx G (1994) Polymer 35:1051
137. De Juana R, Cortazar M (1993) Macromolecules 26:1170
138. De Juana R, Hernandez R, Pena JJ, Santamaria A, Cortazar M (1994) Macromolecules 27:6980
139. De Juana R, Jauregui A, Calahorra E, Cortazar M (1996) Polymer 37:3339
140. Cebe P (1988) Polymer Composites 9:271
141. Ozawa T (1971) Polymer 12:150
142. Ziabiki A (1967) Appl Polym Symp 6:1
143. Ziabiki A (1974) Coll Polym Sci 6:252
144. Watanabe T, Sumi Y, Fujiwara Y, Nishi T (1983) Polym Prepr Japan 31:886
145. Nojima S, Terashima Y, Ashida T (1986) Polymer 27:1007
146. Tanaka H, Nishi T (1989) Phys Rev A 39:783
147. Vaidya MM, Levon K, Pearce EM (1995) J Polym Sci B: Polym Phys 33:2093
148. Defieuw G, Groeninckx G, Reynaers H (1989) Polymer 30:2158
149. Vanneste M, Groeninckx G (1995) Polymer 36:4253
150. Clendening RA, Potts JE (1975) US Pat 3,929,937
151. Iwamoto A, Tokiwa Y (1993) Kobunshi Ronbushu 50:789
152. Iwamoto A, Tokiwa Y (1994) J Appl Polym Sci 52:1357
153. Kang SH, Han S, Jeon IN, Yoon HG, Moon TJ (1996) Polymer (Korea) 20:1080
154. Guo Q, Zheng S, Li J, Mi Y (1997) J Polym Sci A: Polym Chem 35:211
155. Chen H-L, Liaw D-J, Tsai J-S, Shyu J-S, Yang J-M (1996) Polym J 28:976
156. Taesler C, Kricheldorf HR, Petermann J (1994) J Mat Sci 29:3017
157. Kricheldorf HR, Wahlen LH, Friedrich C, Menke TJ (1997) Macromolecules 30:2642
158. Lu X, Weiss RA, Hsiao BS, Wu DQ, Li YJ, Chu B (1995) Polym Prepr Amer Chem Soc Polym Div 33(2):589
159. Eastmond GC, Phillips DG (1979) Polymer 20:1501
160. Nojima S, Wang D, Ashida T (1991) Polym J 23:1473
161. Nojima S, Takahashi Y, Ashida T (1995) Polymer 36:2853
162. Nojima S, Kuroda M, Sasaki S (1997) Polymer J 29:642
163. Tanaka H, Hasegawa H, Hashimoto T (1991) Macromolecules 24:240
164. Larsson H, Bertilsson H (1996) Polym Networks Blends 6:99
165. Callaghan TA, Takakuwa K, Paul DR, Pawda AR (1993) Polymer 34:3796
166. Christiansen WH, Paul DR, Barlow JW (1987) J Appl Polym Sci 34:537
167. Robeson LM, Hale WF, Merriam CN (1981) Macromolecules 14:1644
168. Uriarte C, Eguiazabal JI, Llanos M, Iribarrn JI, Iruin JJ (1987) Macromolecules 20:3038
169. Qipeng G (1990) Eur Polym J 26:1329
170. Guo Q, Zheng H (1999) Polymer 40:637
171. Tsukahara Y, Yonemura T, Hashim AS, Kohjiya S, Kaeriyama K (1996) J Mater Chem 6:1865
172. Messersmith PB, Giannelis EP (1995) J Polym Sci A: Polym Chem 33:1047
173. Jimenez G, Ogata N, Kawai H, Ogihara T (1997) J App Polym Sci 64:4179
174. Tian D, Dubois P, Jerome R (1996) Polymer 37:3983
175. Tian D, Dubois P, Jerome R (1997) J Polym Sci A: Polym Chem 35:2295
176. Tian D, Blacher S, Dubois P, Jerome R (1998) Polymer 39:855
177. Tian D, Dubois P, Grandfils C, Jerome R, Viville P, Lazzaroni R, Bredas J-L, Leprince P (1997) Chem Mater 9:871

Editor: Prof. U.W. Suter
Received: March 1999

Author Index Volumes 101–149

Author Index Volumes 1–100 see Volume 100

Subject Index